Climate Change

American and Comparative Environmental Policy
Sheldon Kamieniecki and Michael E. Kraft, series editors

For a complete list of books in the series, please see the back of the book.

Climate Change

What It Means for Us, Our Children, and
Our Grandchildren

Second edition

Edited by Joseph F. C. DiMento and
Pamela Doughman

The MIT Press
Cambridge, Massachusetts
London, England

MIT Press books may be purchased at special quantity discounts for business or sales promotional use. For information, please email special_sales@ mitpress.mit.edu.

This book was set in Sabon by the MIT Press. Printed and bound in the United States of America.

Pamela Doughman is an Energy Specialist with the California Energy Commission. Her contributions to this book were made in an individual capacity and not as an employee of the California Energy Commission. The opinions, conclusions, and findings expressed in her work in this book are hers alone, and do not necessarily reflect the official position, policies, or opinions of the California Energy Commission or the State of California. The California Energy Commission has not reviewed or approved Pamela Doughman's work for this book.

Library of Congress Cataloging-in-Publication Data

Climate change : what it means for us, our children, and our grandchildren / edited by Joseph F. C. DiMento and Pamela Doughman. — Second edition.
    p. cm.
Includes bibliographical references and index.
ISBN 978-0-262-52587-9 (pbk. : alk. paper)
1. Climatic changes—History. 2. Climatic changes—Social aspects.
3. Communication in science. I. DiMento, Joseph F.
QC903.C56354   2014
577.2'2—dc23
2013036291

10 9 8 7 6 5 4 3 2 1

Contents

Series Foreword

The debate over climate change continues, and reconciling diverse and conflicting interests remains as challenging as it has been for the past decade. The Intergovernmental Panel on Climate Change warns that the risks and costs of climate change grow in severity with greater levels of greenhouse gas emissions. Many US and international environmental groups have allocated substantial time and resources to lobbying political leaders in an effort to control greenhouse gas emissions. Most of their policy proposals, if adopted, will require substantial expenditures by business. Both fossil fuel producers and fossil fuel consumers, including average citizens and a wide variety of small and large companies, will face high costs for even a moderately aggressive program to curtail greenhouse gas emissions. In addition, insurance companies remain concerned about significant claims related to increasingly severe weather (e.g., more powerful hurricanes and tornadoes).

For all these reasons, scientists are under pressure to produce research findings that clearly point to the particular causes of the problem, its scope and severity, how much must be done, and how quickly action must be taken. Additional research on climate change will require significant government funding and will likely take considerable time because of the complexity of the issues.

Meanwhile, American politicians and policymakers are caught in the middle of the contentious debate and must keep a vigilant eye on the results of scientific studies and public opinion. Although public support for action to address climate change is high, the public is likely to reject required and significant lifestyle changes unless the danger is perceived to be real and immediate action is needed.

Like the excellent first edition, the second edition of this edited volume introduces readers to the science, politics, and policy of the climate change debate. Its main goal is to educate students and members of the broad public about the scientific, political, and communication issues concerning climate change by providing balanced and well-documented information and observations about the problem. This is done in response to the efforts by fossil fuel producers and consumers to distort the discourse surrounding climate change and persuade the public that a problem does not exist. The book assumes the following: that the public wants to know more about climate change, that understanding climate change is not easy, and that it does not have to be that way. The various chapters analyze the disconnection between the scientific community's assessment of the importance of the problem and the inaction of some governments, principally the US national government. The book also discusses actions under way at the state and local levels to reduce greenhouse gas emissions and increase resilience to climate change. As readers will see, the various contributors to this edited volume provide a clear, unbiased, and coherent presentation of the various elements associated with the debate over climate change.

As in the first edition, the book begins with a primer on global climate change. It reports on the current level of scientific understanding regarding why the earth is warming and what is causing it. Then, the book discusses the impacts of climate change on the world, on certain regions, and on individual nations. Following these science-oriented chapters, other contributors explore global responses in both the public and private sectors. The media, of course, have played a key role in communicating to the public information about climate change, and one eminent journalist suggests ways in which the media can do a much better job educating the public about this problem. This is followed by an assessment of the impact of climate change on the most vulnerable societies in developing countries as well as risks to the United States. The last chapter pushes readers to the next steps in thinking about climate change and the need to address the problem in light of existing scientific knowledge. Clearly, the longer government postpones action, the worse the climate change problem will become, and the more costly it will be to correct it, assuming it is not too late by then. Both students and average citizens will find the ideas conveyed in this book easy to comprehend, and they are certain to come away with a basic understanding of the climate change controversy.

The second edition of Joseph DiMento and Pamela Doughman's book benefits from considerable new scientific discovery and theory and brings the reader to an understanding of climate science through this point in time. The second edition describes major actions taken by policymakers in light of increased knowledge and understanding about alternative regulatory tools, including the use of taxes and markets in ways that were not known at the time of the first edition. Furthermore, the current volume adds New York State as a case study, and it gives greater attention to developing nations, including China. It provides more assessment of the effectiveness of alternative approaches, additional analyses of the negative effects of climate change on the environment, and an evaluation of the important role played by new social media in the climate change debate. Overall, the second edition offers a significant update of the facts, figures, data, events, politics, and policies concerning climate change reported in the first edition.

The book illustrates well our purpose in the MIT Press series in American and Comparative Environmental Policy. We encourage work that examines a broad range of environmental policy issues. We are particularly interested in volumes that incorporate interdisciplinary research and focus on the links between public policy and environmental problems and issues, both within the United States and in cross-national settings. We welcome contributions that analyze the policy dimensions of relationships between humans and the environment from either a theoretical or empirical perspective.

At a time when environmental policies are increasingly seen as controversial and new and alternative approaches are being widely implemented, we especially encourage studies that assess policy successes and failures, evaluate new institutional arrangements and policy tools, and clarify new directions for environmental politics and policy. The books in this series are written for a wide audience that includes academics, policymakers, environmental scientists and professionals, business and labor leaders, environmental activists, and students concerned with environmental issues. We hope they contribute to public understanding of environmental problems, issues, and policies of concern today and also suggest promising actions for the future.

Sheldon Kamieniecki, University of California, Santa Cruz
and Michael E. Kraft, University of Wisconsin, Green Bay
Co-editors, American and Comparative Environmental Policy
Series

Acknowledgments

This work came about from the collective contributions of a number of people to whom we owe significant thanks. The original idea for the book evolved from a program of the Newkirk Center for Science and Society at the University of California, Irvine (UCI). The Center supported book production, including of the second edition, throughout its many stages. The Center is generously funded by a major gift from Martha and James Newkirk, who also supplied supplemental funding specifically for this work.

We are responsible for the analysis, but it is a more complete treatment thanks to the serious, substantive, and detailed comments that the anonymous MIT reviewers supplied.

Additional funding for the work represented in this book came from UCI's School of Law, Division of Research and Graduate Studies, and School of Social Ecology; the Research Group in International Environmental Cooperation, the Center for Global Peace and Conflict Studies; the Canadian Consulate, Los Angeles; the University of California Institute for Global Conflict and Cooperation; and the California Climate Change Registry. Mrs. Joan Irvine Smith has generously assisted in the funding of dissemination of our results.

To a large number of scientists and analysts we are particularly indebted for comments, advice, and corrections. Most notable are Professors Susan Trumbore and Eric Salzman, Earth System Science, UCI, and James Fleming, Colby College.

The second edition of *Climate Change* benefits from the invaluable research assistance and editing excellence of Dr. Suzanne Levesque.

Clay Morgan of the MIT Press made it a pleasure to produce this work and showed wonderful patience with the logistics of an edited volume with several authors. Also with the MIT Press, many thanks to Meagan Stacey for assistance in production of the first edition. For the second edition, many thanks to Miranda Martin and Virginia Crossman for production assistance, Kristie Reilly for copyediting, Sheila Bodell for the index, and Erin Hasley for the cover design. Sheldon Kamieniecki and Michael E. Kraft, editors of the MIT Press American and Comparative Environmental Policy series, were supportive through the review process and provided helpful insights. The editing skills of Deborah Cantor-Adams and Rosemary Winfield helped us to realize our goal of clearly communicating a complex subject. Production of *Climate Change* was kept on time and professional through the extra efforts of Marlene Dyce, Elizabeth Eastin, T. J. Fudge, Erin Umberg, and Shyla Raghav of the Newkirk Center for Science and Society.

We are grateful to the American Association for the Advancement of Science and the History of Science Society for supporting the George Sarton Memorial Lecture that led to Professor Oreskes' chapter 4, to Myles Allen and David Stainforth for permission to reproduce the figure from climate prediction.net that appears in chapter 4, and to Erik Conway,

Alison MacFarlane, and Leonard Smith for comments on early versions of that chapter.

Many family members and friends, knowing our objectives, have asked us often about climate change, and we have tried to be responsive to their helpful questions. Most notable were the probing comments of Dr. Deborah Newquist, Dr. Louis Di-Mento, Dr. Lawrence Steinberg, and the late Dr. Duncan Luce.

1

Introduction: Making Climate Change Understandable

Joseph F. C. DiMento and Pamela Doughman

Over the centuries, humans have tried to change the weather. People have prayed, danced, seeded clouds, and used other strategies to get more rain, stop the rain, decrease the heat, and warm things up a bit. Seldom have we deliberately tried to change *climate*—the average weather conditions over an extended period of time—but we have unintentionally changed climate historically, and we are changing it today.

This book draws on the vast knowledge of earth system science to explore those changes in climate, including important changes linked to the level of what are known as *greenhouse gases*—the 3 percent of the gases in the earth's atmosphere that help to warm the planet and otherwise disrupt the climate. It addresses how such changes may affect us, our children, and our grandchildren—globally and locally. It then explains what is being done and what can be done to manage these changes for a more livable planet.

Global climate change is a major societal issue that many citizens do not understand, do not take seriously, or do not consider to be a major public policy concern. Bill McKibben noted in 2003 that people think of climate change as they do of the trade deficit and violence on television—"as a marginal concern for them ... if a concern at all" (McKibben 2003). That summary remains largely true, although some citizens now understand that helpful actions are possible, while others are getting increasingly worried or feeling helpless.

Yet the scientific community, with the exception of a few contrarians, sees climate as one of the major challenges facing society in the coming decades. Climate disruption is described as serious and unequivocally linked by scientists to human impact (Millennium Alliance for Humanity and the Biosphere 2013). And policy scientists recognize that better environmental management is feasible. This book aims to make climate change understandable to the public—from its historical roots to future available responses.

Scientists and policymakers face some skeptical populations. Views on major questions of public policy are often deeply ingrained and are held like religious beliefs. When citizens conclude they are being asked to change their daily behaviors (drive and air condition less, eat less meat) they may respond in doctrinaire ways—looking to information and sources to reinforce their views rather than challenge them.

Disbelief comes from other sources too. We have a number of friends who respond to assertions that climate change is a problem by saying that such concerns are part of a "disaster strategy." They feel that scientists and policymakers articulate dire environmental futures because it is in their professional interest to do so. One developer friend put it this way: "I remember years ago, during a short downturn in the fish catch in Upper Newport Bay, California, we were told that the situation was hopeless and that the future was one of us being fished out. This year, like others back and forth, we have had great catches. I just don't believe some of these scare scenarios."

The percentage of people in the United States who "believe in climate change" or believe that changes are linked to human behavior oscillates. In 2013, a survey conducted by Pew Research Center found that most Americans (about 7 in 10) believe the global average temperature is getting warmer; the survey found about four in ten said human activity was the main cause. Pew found similar results in 2008, higher results in June 2006 (79 percent and 47 percent, respectively), and lower results in October 2009 (57 percent and 36 percent; Desilver 2013). In another survey conducted in 2013, Pew found people in other countries are more concerned than Americans about the threat of global climate change for their country (Drake 2013).

We know others who conclude, "There's nothing I can do," in a type of learned helplessness. The population was 2.5 billion in 1950, when some of them were born; it will be 9.6 billion in 2050, and it is already 7.2 billion, with most new growth occurring in countries that are only now turning to industrialization and what their leaders consider the essential use of fossil fuels. How does our choice of an automobile type make any difference when developing nations are adding coal-fired power plants by the week? China and India, in this view, are no longer third-world countries that should be immune from controls on greenhouse gas emissions. This view does not reflect per capita contributions to greenhouse gases, on which the United States, for example, continues to be worse than most other countries. Inaction on climate change among those who are informed and concerned is a complex but important barrier to social change (Norgaard 2011, 3).

The global economic recession since the publication of our first edition may set the context for the thinking of still others. For some, immediate needs are so pressing they demand full attention, making consideration of future impacts difficult. Some may be so strapped, they think, "After we make it through today, we'll have time to worry about tomorrow."

Others feel that the planet's climate has always gone through change and that we are simply experiencing another phase. This is scientifically accurate, but what is happening now is different:

> The difference now, in a nutshell, is civilization, which developed during a period of exceptional climatic stability over the last 7,000 years … . [H]uman society has … become finely adapted to the current climate, so much so that a mere three-foot increase in sea level (small by the standards of geologically recent changes) would displace around 100 million people. Agriculture and animal husbandry are also tuned to the present climate, so that comparatively small shifts in precipitation and temperature can exert considerable pressure on government and social systems whose failures to respond could lead to famine, disease, mass emigrations, and political instability. (Emanuel 2012, 54–55)

Some look to last week's cold spell and generalize from that. And others argue that it is in the professional interest of some actors to continue business as usual. An article in *Rolling Stone* recommended divestment of fossil fuel companies as a strategy to raise political pressure to leave 80 percent of known reserves, on the order of $20 trillion in value, in the ground. If burned, those reserves would produce profit for fossil fuel companies and their investors. However, they would also produce five times the maximum safe global emission level for 2050 (McKibben 2012).

Some governments still have not been persuaded to take action to address climate change, even though the scientific community reports that inaction is likely to lead to quite difficult and costly consequences. But not everyone downplays the threat of global climate change. Countless individuals, many industries, thousands of cities, and many countries are taking action to address climate change, responding to the science of climate change in ways that improve conditions for us, our children, and our grandchildren.

And some have begun to appreciate that "a clean-energy economy would break the world's dependence on a small set of security-fragile regimes, stop funding nondemocratic states, free up resources for other forms of protection, and allow us to deploy money and human energy toward safer and more productive uses" (Wapner 2010, 192).

This book assumes three things: the public would like to understand climate change better, and understanding climate change is not easy—but it doesn't have to be that way.

Obstacles to Understanding Climate Change

Climate change can be difficult to understand for four key reasons, ranging from a public perception of lack of scientific consensus to the complexity of the science itself. Below, we explore each of these barriers.

Some people believe scientists lack consensus on the human contribution to climate change

A public perception of lack of consensus has been powered, in part, by strong statements made by politicians who have selectively parsed the words of mainstream scientists and used the conclusions of those who are outliers on the subject. One outspoken United States senator regularly refers to global climate change as the largest hoax ever perpetrated on the American people. In part, the perception of a weak scientific base also comes from popular culture, such as fictional works like Michael Crichton's 2004 novel *State of Fear*. To write a compelling piece of fiction, he used footnotes and other gimmicks to give the impression that science is being abused by climate-change investigators.

Amplifying uncertainties about climate change has at times been consistent with the national political interest of the United States and other nations highly responsible for greenhouse gas emissions, and is similar to the way some environmental groups have minimized the same uncertainties, at times causing a backlash. Outcomes that are "likely" can easily convert to statements of either fact or uncertainty. In reality, the link between human activities and climate disruption is widely accepted by scientists who work in the many fields that study climate and the factors associated with its change.

The conclusion that a scientific consensus on climate change does not exist also derives from the way climate change has been treated in the media. Ethical journalists are committed to presenting controversial subjects fairly and to searching out contrarian viewpoints. But readers of newspaper and magazine articles, viewers of television news programs, and Internet surfers may be left with an impression that climate change is both a major worry and nothing to be concerned about.

The media also tend not to provide in-depth coverage of environmental issues, especially those that do not have immediate, dramatic effects. Responsible ongoing coverage will seldom impress a public bombarded with more easily graspable stories. Occasionally, climate change is a front-page news item, as with the case of hurricane damage, but later coverage might question whether the original story was merited or even accurate. When accurate science stories reach the popular media's front pages or major segments, they often are packaged in a way that seems exaggerated. There are exceptions, such as the weekly *Science* section of the *New York Times* and some blogs, but even major newspapers seem compelled to act like they are striving for balance or avoiding bias by reporting positions of outliers in ways that would seem shocking if they were addressing a topic about which the general public knows more.

A perception of nonconsensus on climate change also results from the nature of the science, which is based both on computer models and also on actual tests, measurements, and completed studies in the field. Computer models are powerful tools that make assumptions that are not always considered credible by critics, and even mainline scientists agree that the models need refinement. For example, one of our authors has written that the models need to reflect an improved

understanding of the aerosols spewed by smokestacks, unfiltered tailpipes and volcanoes. They were once presumed only to have a cooling influence. Now, however, aerosols are known to cause both cooling and warming, depending on their color and composition and how they affect clouds, whose properties are slowly being incorporated in the simulations. (Revkin 2004)

However, even since the publication of our first edition, models have become more sophisticated. In addition, the many fields that educate us about climate change are maturing and, to the reasonably open-minded observer, becoming more convincing.

Based on a growing body of observations, field research, ice core drilling, and increasingly detailed computer models, a majority of scientists share the conclusion that climate change is real, serious, and human-induced. Clearly measuring and communicating how much, within convincing and influential ranges, is human-caused and how much results from "natural variability" is an important step in educating the public and perhaps influencing policymakers. The US National Academy of Sciences (NAS) reports, "Carbon dioxide emitted from the burning of fossil fuel is presently the largest single climate forcing agent, accounting for more than half of the total positive forcing since 1750" (NAS 2011). This does not include effects of deforestation or other human activities on climate change—just the burning of fossil fuels.

In recent years it has been difficult to deny some form of climate change, since even the most skeptical have witnessed climate-caused migration of people and plants and animals, including the iconic polar bear; the opening of the Arctic to commercial shipping; the legal response almost worldwide in new laws and lawsuits; Fortune 500 companies' mandated reporting to shareholders regarding climate-linked vulnerabilities; and the decisions of insurers, including the most globally significant, to exclude from coverage some damages connected with changes in climate.

Scientists work with probabilities, risks, ranges, uncertainties, and scenarios—approaches foreign to many people

People often learn of scientific findings from experts who are not trained in communication or are trained to communicate only with their peers. Even when risk is defined, it can be confusing to those who have not studied the social sciences, and probabilities are difficult for many "laypeople" to grasp. It is common to overestimate the chances of one type of danger, such as flying in a commercial jet, and underestimate those of others, such as speeding or texting while driving (Blastland and Spiegelhalter 2013). And people may hold different views on self-imposed risks compared to risks imposed on them by others. Furthermore, models that earth system scientists, atmospheric chemists, and others consider simple are not easy for people outside the field to follow.

While we may not all be "climate change idiots," as one journalist termed it, many people think in ways that make understanding the science of climate change difficult. There is a new field of climate psychology that explains why many cannot grasp

the more abstract, global dangers posed by climate change. ... We have trouble imagining a future drastically different from the present. We block out complex problems that lack simple solutions. We dislike delayed benefits and so are reluctant to sacrifice today for future gains. And we find it harder to confront problems that creep up on us than emergencies that hit quickly.

"You almost couldn't design a problem that is a worse fit with our underlying psychology," says Anthony Leiserowitz, director of the Yale Project on Climate Change Communication. (Gardiner 2012)

As Dan M. Kahan of Yale and others have noted, we look to information that is available and that reinforces what we already believe (Gardiner 2012). If we believe that climate change is caused by changes in sunspots or that the changes we are experiencing are simply the repeat of climate cycles that the earth has experienced for thousands of years, we will likely continue to hold on to such positions even in the face of National Academy of Sciences or other scientific studies that refute these notions.

The vocabulary, science, and policies of climate change are complex

The climate field is peppered with terms such as *sinks, forcing, secondary effects, adaptive capacity, greenhouse gas equivalent, albedo, carbon cycle, integrated assessment, no-regrets policy, cap and trade, net primary production, joint development*, and *clean-development mechanism*, as well as numerous acronyms. In fact, there are almost three hundred terms in the glossary of one assessment report from one working group of the Intergovernmental Panel on Climate Change (IPCC). The IPCC, created in 1988, is the main international body established by the World Meteorological Organization and the United Nations Environment Program to assess climate-change science and provide advice to the international community. In this book, we use everyday language terms or define a term that is not common when it is first used. We use the terms *anthropogenic* and *human* interchangeably.

Related to the complex vocabulary and tools of climate science—and in some ways quite trivial, but a point we make in our conversations with international scientists: much of the science of climate change gets reported in the metric system, which is foreign to many Americans. A change of 5.8° Celsius may not seem of concern to a New Yorker, but an increase of 10.5° Fahrenheit seems large—although they are equivalent. And 40° Celsius is 104° Fahrenheit. It is noticeably warmer today than when Henry David Thoreau wrote about natural and environmental history in the early to mid-1800s.

The environmental and social effects of climate change are not discrete or evenly distributed

The effects of climate change do not cluster in ways that can be clearly linked by the nonscientist (and in some cases, even by scientists) to climate-change dynamics. This particular drought, that remarkable cold spell in Moscow, or a recent hurricane may have occurred regardless of whether overall climate characteristics are changing. The environmental and social impacts are and will be unevenly distributed, even within countries. So it is not uncommon that people living in different regions of a country differ in their views of what, if anything, is going on and what needs to be done. If the climate seems milder in upstate New York or the open shipping-lane season of the Arctic is longer, why should we be concerned? There are even benefits linked to these changes. Further, climate change will create wealth for some: not only the existing fossil fuel industries that profit from business as usual, but also entrepreneurs, industries, and institutions, which will see rising market demand for new technologies, processes, and institutions. Finally, even among some parts of the populace dedicated to improving the global environment and global environmental public health, there is a sense that we should be working on, and dedicating resources to, more immediate and demanding challenges—including those with clearly achievable solutions (such as wiping out malaria, polio, and tuberculosis), thus promoting environmental justice in more direct and concrete ways.

Preview of the Book

This book responds to these challenges. The following chapters bring the how, what, and why of climate change from the laboratory and think tank into the living room. *Climate Change: What It Means for Us, Our Children, and Our Grandchildren* summarizes in understandable terms what science knows about climate change and addresses how that knowledge has been used and can be turned to action by government, business, and everyday citizens. This book also recommends ways to further the public's understanding of this complex international environmental challenge and affect public opinion in ways that may drive policy and actions.

The book first offers a primer on global climate change. Chapter 2 explains the nuts and bolts of climate and the greenhouse effect, as well as the discovery of their interaction. Next, in chapter 3, we summarize the effects of climate change on the world, regions, and states. Here we describe how people, plants, animals, crops, and the natural environment are all affected by climate change. Adding a science historian's perspective, in chapter 4 Naomi Oreskes explains the nature of consensus in the climate-change debate and asks how we know that we're not wrong—and whether the contrarians and merchants of doubt might yet carry the day. In this chapter, as in others, we again encounter the mammoth amount of work compiled by national and international scientific bodies, one of which, the IPCC, has undertaken five comprehensive assessments of the state of the world's climate.

Following these science-based chapters, in chapter 5 we explore world responses from the public and private sectors. How have international scientific and legal organizations reacted? What has been the US position, and how has it changed, covering events through the summer 2013 announcement by the Obama administration that it will employ its regulatory authority to set standards for existing coal-fired power plants? How does the US stance differ from those of countries in Europe and Asia, including Japan, China, and India—and from small island states whose very existence is threatened by climate disruption? We describe how national positions have moved in fits and starts through the Kyoto and pre-Kyoto periods (referring to the major international protocol that is part of the United Nations' efforts to address climate change). How have states and businesses, large and small, responded? Can New York, California, and other states, as well as BP (formerly known as British Petroleum), General Electric (GE), and other major companies mitigate climate-change effects? Will market thinking lead the way to either mitigation of greenhouse gases or adaptation to changes? Will we be able to engineer our way out of some of the problems, reflecting heat back into space, or fertilizing the oceans to slow the process of acidification?

The book then turns squarely to the question of how climate science is communicated. In chapter 6, science writer Andrew C. Revkin shares his experiences and ideas for improvement. An important factor in societal decisions about action and nonaction is the manner in which scientific information is understood and communicated. Chapter 6 discusses changes over time in pressures facing reporters, including in this period of massive use of social media, which has exploded since our first edition.

In chapter 7, Richard Matthew addresses the effect of climate change on vulnerable populations, especially in developing countries, but also touching on areas at risk in the United States. Here and elsewhere in the book, we address environmental justice as climate change affects it. What is fair in attributing cause? What is just in crafting solutions? Our concluding chapter takes readers to next steps in thinking about climate change and in acting on the science: we discuss the costs and benefits (often highly contested) of actions of various sorts, policy responses, and the roles that are appropriate for governments, businesses, and citizens. We conclude that much can be done to manage climate change in the context of ever-growing challenges related to population growth and industrialization. Although new knowledge in many fields continues to improve our understanding of climate change, we have enough information to act now, and an obligation to future generations to avoid delay.

References

Blastland, Michael, and David Spiegelhalter. 2013. *The Norm Chronicles: Stories and Numbers about Danger*. London: Profile Books.

Desilver, Drew. 2013. Most Americans say global warming is real, but opinions split on why. *FactTank: News in the Numbers*. Pew Research

Center. June 6. http://www.pewresearch.org/fact-tank/2013/06/06/most-americans-say-global-warming-is-real-but-opinions-split-on-why/, accessed July 5, 2013.

Drake, Bruce. 2013. Most Americans believe climate change is real, but fewer see it as a threat. *FactTank: News in the Numbers.* Pew Research Center. June 27. http://www.pewresearch.org/fact-tank/2013/06/27/most-americans-believe-climate-change-is-real-but-fewer-see-it-as-a-threat/, accessed July 5, 2013.

Emanuel, Kerry. 2012. *What We Know About Climate Change.* Cambridge, MA: MIT Press.

Gardiner, Beth. 2012. We're All Climate-Change Idiots. *New York Times,* July 21. http://www.nytimes.com/2012/07/22/opinion/sunday/were-all-climate-change-idiots.html?_r=0, accessed July 3, 2013.

McKibben, Bill. 2003. Worried? Us? *Granta* 83 (Fall): 7–12.

McKibben, Bill. 2012. Global warming's terrifying new math: Three simple numbers that add up to global catastrophe—and that make clear who the real enemy is. *Rolling Stone Magazine.* August.

Millennium Alliance for Humanity and the Biosphere. 2013. Scientific Consensus on Maintaining Humanity's Life Support Systems in the 21st Century.

Norgaard, Kari Marie. 2011. *Living in Denial: Climate Change, Emotions, and Everyday Life.* Cambridge, MA: MIT Press.

Revkin, Andrew C. 2004. Computers add sophistication, but don't resolve climate debate. *New York Times,* August 31, D-3.

US National Academy of Sciences. 2011. What is Climate Forcing? Koshland Science Museum. Slide 4. As cited by EPA 2013 in Radiative Forcing, Causes of Climate Change. http://www.epa.gov/climatechange/science/causes.html, accessed July 5, 2013.

Wapner, Paul. 2010. *Living through the End of Nature: The Future of American Environmentalism.* Cambridge, MA: The MIT Press.

2

A Primer on Global Climate-Change Science

John T. Abatzoglou, Joseph F. C. DiMento, Pamela Doughman, and Stefano Nespor

Climate change is occurring, is caused largely by human activities, and poses significant risks for—and in many cases is already affecting—a broad range of human and natural systems.

—National Academy of Sciences (2010, 3)

The earth's climate system includes a series of checks and balances that, in the past, have worked together to maintain a stable climate. However, mounting evidence suggests that this balancing act has been tested over the last century and a half. Multiple indicators—including increasing air and ocean temperatures, increasing sea level, and retreating glaciers and sea ice—all point to a warming planet.

In this chapter, we present a primer on climate science and climate change. Climate, by its place-based definition, is an intuitive concept—we know it's usually warmer in Honolulu, Hawaii, than in Fairbanks, Alaska. However, climate science is composed of an intricate system that involves the interaction of air, water, ice, plants and animals, soils, and the solid earth. The complex nature of the earth's climate system presents a difficult scientific challenge, as it involves numerous interacting scientific disciplines. The interaction of multiple disciplines encourages scientific advances that speed up the evolution of climate science. This chapter summarizes several key elements of climate science to give readers a working background on the topic.

Climate and Weather

It is important to distinguish between weather and climate. Robert Heinlein (1978) provides us with a good starting point: "Climate is what you expect; weather is what you get." *Weather* refers to meteorological conditions at a particular time and place. The commonly used conversation starter: "How's the weather?" can be answered through a description of current outside conditions, including the temperature and the presence or absence of rainfall, cloud cover, humidity, and wind.

Weather forecasts are frequently used in a decision-making context and have been shown to be economically beneficial (Lazo et al. 2009). Weather forecasts are used to guide everything from the rather trivial issue of deciding whether to carry an umbrella or pack a coat to nontrivial issues associated with preventing or lessening weather-related damages or disasters. Weather predictions involve gathering observations and using mathematical models based on the principles of physics to describe probable atmospheric conditions for the next few days. The accuracy of weather forecasts has improved over the past couple of decades due to advances in observational techniques and modeling, and are extensively used to help us make informed decisions that save money and lives.

Forecast skill drops off after about ten days, chiefly because minute inaccuracies in observations incorporated into forecast models increase over time. This is *chaos theory*, often referred to as the "butterfly effect," after the idea that a butterfly flapping its wings in Brazil could create a localized swirl in atmospheric flow that could cascade and impact weather conditions at larger scales (such as creating a tornado in Texas). Of course, we are unable to monitor the movement of each of the world's butterflies, and thus forecast models tend to develop errors at longer timescales.

Climate refers to both the average and range of weather conditions that occur over an extended period of time (months, years, or centuries). The saying "climate is what we expect" therefore includes both a single average value and a range of outcomes taken from historic conditions. In this way, climate is composed of numerous samples of weather. Metaphorically, weather is one's mood whereas climate is one's personality, and weather is what you had for breakfast whereas climate is what is in your pantry. While climate is considered to summarize weather across a historic time period, it is important to note that changes in climate are noted when weather conditions exceed previously established ranges of variability. Although *weather* and *climate* are distinctly separate terms, just as there are daily fluctuations in atmospheric conditions (weather), there are also long-term fluctuations in climate. Deviations from established climate records can occur on a monthly or seasonal basis; for example, a wet winter in California is made up of a greater frequency or intensity of wet weather. They can also occur across longer timescales; for example, warming observed during spring across western North America over the last half-century is made of up warmer daytime high and low temperatures.

While we may not experience climate on a daily basis, as we do weather, it plays a strong role in defining human and environmental geography. Climate enables humans to settle in a specific area by defining natural resources such as water and agricultural potential within a local area. Climate influences human culture, from the articles of clothing we wear and the spices we use to the recreational activities we enjoy. While climate may be what we expect and depend upon, variations in climate disrupt the processes, such as food production, water availability, and energy demands, that societies are built upon. Variability in climate has acted, and continues to act, as a stressor on society, and economic losses from climate impacts persist despite modern advances and attempts to "climate-proof" resources and infrastructure. As we have become more aware of hazards associated with climatic variability, a greater emphasis has been placed on integrating seasonal climate forecasts into decision making, similar to our use of weather forecasts.

Earth's Climate System

The earth's climate system can be thought of as an elaborate balancing act of energy, water, and chemistry through the atmosphere, oceans, ice masses, biosphere, and land surface. These internal components of the climate system influence the fate of energy emitted by the sun and received by the earth. Solar energy, or solar *radiation*, serves as the impetus of energy in the earth's climate system. Roughly 30 percent of the solar radiation directed toward the earth is reflected back to space by bright, reflective surfaces, including snow cover, sand, and clouds. This is the same phenomenon that keeps a white car much cooler than a black car on a hot day.

Just as the sun emits energy, so does each object whose temperature is above absolute zero, including the Earth's surface. Absolute zero is the temperature at which, in theory, particles have no energy (Merall 2013). Emitted energy, or *radiation*, from both the sun and the earth travels in the form of waves that are similar to the waves moving across the surface of a pond.

However, the energy emitted by the sun and earth is quite different. Each emits radiation at distinctly different *wavelengths* (the distance between adjacent crests in the wave) and temperatures. While the hot sun emits energy at short wavelengths (referred to as *shortwave radiation,* including *visible light*), the much cooler earth emits radiation at longer wavelengths (referred to as *longwave,* or *thermal, radiation*).

For the earth to maintain a stable temperature, there must be a balance between the amount of radiation absorbed by the earth and the amount of energy emitted from the earth back to space. According to simple energy-balance calculations, the average temperature of the earth's surface in the absence of an atmosphere should be –18°C (0°F). Fortunately, the earth has an atmosphere that traps much of the thermal radiation emitted by the earth's surface, but allows most of the solar radiation to pass through. This acts somewhat like a one-way mirror seen in interrogation rooms on crime dramas. Certain trace gases in the earth's atmosphere, called *greenhouse gases,* selectively absorb longer wavelengths of energy emitted by the earth, heating the surrounding atmosphere. This energy is ultimately reflected back to the earth's surface. As a result, the earth's surface has a more difficult time cooling off, and the earth's surface and its lower atmosphere warm significantly.

Greenhouse gases are a natural component of the earth's atmosphere, with water vapor accounting for much of the greenhouse effect. The warming effect of water vapor can be observed during winter nights when the earth's surface has an extended period of time to cool off. A cloudy winter night is typically warmer than a clear winter night. Clouds and water vapor trap the heat radiating from the surface and keep surface temperatures from dropping as much as they would on a clear night. Overall, the natural greenhouse effect allows the average

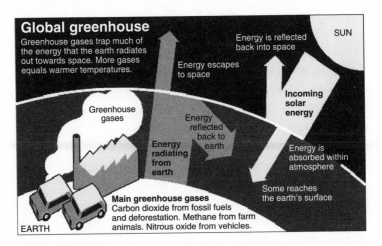

Figure 2.1
Fundamental dynamics of the greenhouse effect.

surface temperature of the earth to warm from a frigid –18°C (0°F) to a more comfortable 15°C (59°F). Thus, the chemical makeup of the atmosphere is crucial in establishing a climate that is hospitable to life (figure 2.1).

Our description of energy balance thus far applies to the earth as a whole. At local scales, the energy balance equation changes. To understand this, consider that solar radiation is much more intense near the equator than near the poles as a result of the curvature of the earth's surface and tilt of the earth's axis. The amount of solar radiation received in the tropics is much larger than that received at the poles. However, the rate at which earth emits energy (radiation) to space does not differ as dramatically from the equator to the poles. As a result, there is a net loss of energy near the poles and a net gain of energy in the tropics. Therefore, the ocean and atmosphere transport excess heat from the tropics to the heat-deficient polar regions via winds and ocean currents. If this transfer did not occur, there would be a rapid cooling of the poles and a dramatic warming of the tropics. Because our atmosphere and oceans redistribute this energy imbalance, much of our earth is livable.

The atmosphere responds to unequal heating of the earth's surface by generating atmospheric motion. *Atmospheric circulation* redistributes heat around the globe in an attempt to create energy balance. Atmospheric circulation responds relatively quickly to radiation. A familiar example of such a response is the sea breeze. During a typical summer day, the land surface heats up much faster than the nearby ocean surface. The warmer air over land becomes less dense and begins to rise. As this air mass rises, the cooler, denser air over the ocean flows inland to replace the rising warm air. As a result, the sea breeze cools inland locations and offsets temperature differences. A similar, more involved process serves to counter global-scale imbalances in heating: intense surface heating of the tropics creates warm buoyant air that rises to the upper troposphere and moves poleward while cold dense air near the poles sinks to the surface and moves toward the equator. This circulation, known as the *Hadley circulation*, redistributes energy around the globe and increases the area of our planet with temperatures hospitable to life.

Oceans are a key component of the climate system. Among the unique properties of water is its ability to store and transport vast quantities of heat. As surface water in the tropics is heated, large-scale ocean currents, driven by atmospheric circulation patterns, transport heat poleward. This process parallels the atmosphere's redistribution of energy across the globe, but it operates on much longer timescales. Consider, for example, the circulation of the North Atlantic basin. The northward-flowing Gulf Stream transports warm water from the Gulf of

Mexico toward northern Europe. It is believed that the Gulf Stream helps to stabilize the mild climates of northwestern Europe (Seager 2006). The ocean also plays a role in determining the chemical composition of the atmosphere due to its ability to take up and release gases important in establishing the earth's climate, such as carbon dioxide, or CO_2.

The *cryosphere* comprises that portion of the planet's water that is locked away as ice, including the Greenland and Antarctic ice sheets, sea ice in the Arctic and Southern Oceans, and all other snow- and ice-covered surfaces. An important property of the cryosphere is its ability to reflect solar radiation. Large ice sheets reflect between 70 and 90 percent of solar radiation, allowing very little radiant energy to warm the surface. If the ice sheet were to grow in extent in response to cooling, a greater amount of solar radiation would be reflected back to space, thereby further cooling the planet. This is an example of a *climate feedback,* where an initial change in one aspect of the dynamic climate system (e.g., slight cooling) interacts with other components of the system and can then alter the magnitude of the original change. The growth of the ice sheet during a glacial period acts as a positive feedback on the climate system. A *positive feedback* involves a coupled system that amplifies the initial change. Conversely, as the ice sheet recedes during warming, more solar radiation reaches the earth's surface, accelerating the warming of the planet and the melting of the ice sheet.

The land surface and the biosphere both affect and are affected by atmospheric temperature and humidity, and can alter the amount of solar radiation reflected back to space. Vegetation also plays a key role in the *carbon cycle*, which is the exchange of carbon among atmosphere, ocean, and land (biosphere included). Plants are active participants in the carbon cycle as they absorb CO_2 through photosynthesis and expel CO_2 through respiration. It is currently thought that plants take up more carbon from the atmosphere than they emit, and are therefore net *sinks* of atmospheric carbon (Le Quere et al. 2009). By contrast, changes in land use—such as deforestation and subsequent burning of forest material—release stored carbon into the atmosphere and are net *sources* of carbon to the atmosphere. In particularly productive ecosystems such as tropical rainforests, deforestation not only provides a source of atmospheric carbon, but also removes a net sink that would otherwise take up atmospheric carbon to build woody biomass through photosynthesis.

The Importance of Greenhouse Gases to Climate

In 1824, the French scientist Joseph Fourier hypothesized that the average temperature of the planet is warmer because of the existence of the earth's atmosphere. He claimed that the warming effect of the atmosphere on the earth's surface was similar to the way a plant warms when it is encased in a house of glass. Fourier called this phenomenon the *greenhouse effect,* and the name stuck.

The composition of the earth's atmosphere governs the climate of the planet and establishes conditions vital for life. The ability of greenhouse gases to warm the surface of the planet depends on four main factors: their efficiency in absorbing heat energy, their concentration and distribution in the atmosphere, and how long they remain in the atmosphere.

Although the atmosphere is primarily composed of nitrogen and oxygen, these gases do not interact with the thermal radiation emitted by the earth. However, greenhouse gases—including carbon dioxide (CO_2), methane (CH_4), nitrous oxide

Box 2.1
Definition of Climate

The word *climate* is derived from the Greek word *klima*, a term that refers to the angle of the sun's rays as they strike the earth's surface. The Greek geographer Ptolemy proposed that changes in tilt according to latitude affected the length of the day and the brightness of the sun—ultimately altering the nature of climate and the viability of life on earth. In addition, geographic differences in the amount of solar radiation over the seasons, along with the position of the continents and mountains, result in regional variations in temperature, wind, and precipitation patterns. The regional climates that are commonly accepted today come from Wildimir Koppen's subdivision of the earth based on temperature, precipitation, and distribution of natural vegetation. Climate ultimately represents the complex web of factors that define the atmospheric conditions in a determined geographical area for a prolonged period of time (years, decades, centuries, and geological eras).

(N_2O), halocarbons, ozone (O_3), and water vapor (H_2O)—are very effective at absorbing thermal radiation. The absorption of energy by air molecules heats the atmosphere, which then reradiates energy back to the surface of the earth. This process prevents the earth from cooling. Another important property of greenhouses gases is that while they are effective at absorbing thermal radiation, they are essentially transparent to solar radiation. Hence, the overall influence of greenhouse gases is to warm the planet to approximately 33°C (59°F), which is remarkable given their seemingly small concentrations in the makeup of the atmosphere.

Two planets closest to the earth offer good examples of how changes in atmospheric composition can lead to changes in surface temperatures. Although Venus is closer to the sun than the earth and thus receives a greater amount of incoming solar radiation, clouds engulf the planet, reflecting nearly 75 percent of the solar radiation. As a result, the solar radiation absorbed by Venus is actually less than that absorbed by earth. However, the thick, carbon dioxide–rich (97 percent CO_2) Venusian atmosphere is highly effective at keeping thermal radiation from escaping to space, resulting in an average surface temperature of 470°C (878°F). In contrast, Mars has a very thin atmosphere with a minimal greenhouse effect. As a result, most of the heat radiated from the surface of Mars escapes to space, and the average surface temperature on Mars is about –60°C (–76°F).

Efficiency of Greenhouse Gases in Absorbing Heat

Scientists estimate the heat-trapping efficiency of different greenhouse gases using an index called *global warming potential* (GWP). This represents the ratio of energy trapped by the earth-atmosphere system for a given mass of a particular gas in comparison with the ability of the same mass of CO_2 to trap energy over a specified period of time. The GWP of CO_2 is defined as 1. By comparison, methane has a GWP of 21, meaning that a given mass of methane can heat the planet twenty-one times as much as the same mass of CO_2. Other greenhouse gases have even larger GWPs. Nitrous oxide and halocarbons have GWPs of 300 and over 5,000, respectively. So although carbon dioxide is notorious for its role in global warming, other, less well-known greenhouse gases also play potent roles in the global warming process.

Quantities of Greenhouse Gases

Scientists can quantify the composition of the atmosphere prior to the historical record by examining other physical records that provide a stand-in for direct measurements. Bubbles of air embedded within ice cores extracted from the Greenland and Antarctic ice sheets reveal a substantial amount of information on past changes in climate. The cores tell us that from about 800,000 years ago until the beginning of the Industrial Revolution in the late 1700s, CO_2 varied from about 180 parts per million (ppm) during glacial periods to about 280 ppm during interglacial periods (Lüthi et al. 2008). During interglacial periods, such as the one we are in today, levels of atmospheric CO_2 have previously been relatively constant, maintained through a balance of carbon exchanges between the atmosphere, biosphere, and oceans. *Anthropogenic*, or human-made, carbon emissions have resulted in an imbalance in the carbon cycle and an accumulation of carbon in the atmosphere. Atmospheric CO_2 levels at Mauna Loa, Hawaii, reached as high as 400 ppm in May 2013 (Mohan 2013). The rate of increase of atmospheric CO_2 since 2000 has a higher average than the rate of increase for each decade going back to the 1960s (Tans and Keeling 2013). The question is: how will these high levels of CO_2 affect our climate?

US emissions of carbon dioxide from fossil fuel combustion in 2011 were about 11 percent above 1990 levels, accounting for about 78 percent of US greenhouse gas emissions weighted by GWP from 1990 to 2011 (EPA 2013a). Likewise, methane and nitrous oxide are naturally occurring gases that have seen their atmospheric concentrations increase due to industrial and agricultural activities and biomass burning. Figure 2.2 shows US greenhouse gas emissions for 2011 by gas. Throughout this book the term *forcing*, or *climate forcing*, refers to a mechanism that alters the amount of energy contained within the climate system. Figure 2.3 shows global radiative forcing by gas for 2011.

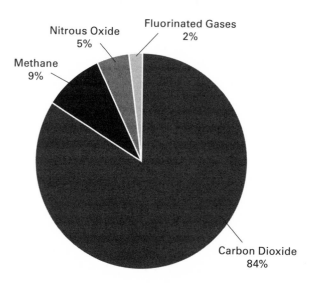

Figure 2.2
United States emissions of greenhouse gases by type in 2011.
Source: US Environmental Protection Agency (2013a). Inventory of
US Greenhouse Gas Emissions and Sinks 1990–2011.

Halocarbons do not exist in nature, but are manufactured
for use in refrigeration units and foaming agents. The most
notorious halocarbons are chlorofluorocarbons (CFCs), which
are potent greenhouse gases that also destroy the ozone layer.
Alternative halocarbon compounds, such as hydrofluorocarbons and perfluorocarbons, were introduced as substitutes for
chlorofluorocarbons. The ozone layer protects life from the
dangerous ultraviolet rays that have detrimental effects on terrestrial and marine life (World Meteorological Organization
2002). Because of the strong scientific evidence linking emissions of CFCs to the depletion of the ozone layer and a global
policy to curtail the usage of CFCs, the chlorine levels in earth's
ozone layer have declined. This success story provides hope

for science-based policy action for other environmental problems the earth faces. However, as a European climate authority notes, "The timing of the ozone layer recovery and in particular the closure of the ozone hole is particularly difficult to forecast since the variability and future evolution of the ozone layer is affected by a number of processes, among them climate change which exerts an influence on atmospheric dynamics and—via temperature changes—on ozone chemistry" (European Space Agency Ozone Climate Change Initiative 2013, 2–3).

Lifetimes of Greenhouse Gases

The atmospheric lifetime of a gas refers to the average amount of time a gas molecule remains in the atmosphere before being

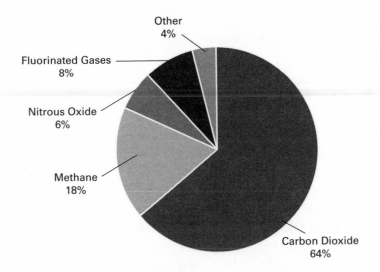

Figure 2.3
Global radiative forcing of all long-lived greenhouse gases in 2011.
Source: US National Oceanic and Atmospheric Administration (2012).

washed out. Water in the atmosphere has a short lifetime of about 10 days. Consequently, it is not well mixed in the atmosphere (as we can observe in the passage of clouds in the sky) and is widely variable across the globe. Other greenhouse gases have long lifetimes of decades to centuries. These gases are well mixed in the atmosphere, because they are transported across the globe by atmospheric circulation. The long lifetimes of greenhouse gases mean they have a cumulative effect on global climate changes. For example, today's emissions of carbon dioxide are likely to remain in the atmosphere for twenty to one hundred years or more.

Industry and Greenhouse Gases

Atmospheric concentrations of greenhouse gases have largely been balanced by the carbon cycle over the life history of the planet. Carbon cycles through the atmosphere via photosynthesis and respiration by land and sea flora, via air-sea exchanges (*fluxes*), and via "slow-turnover" geologic processes. While important for the evolution of the atmosphere, the relatively slow pace of the geologic carbon cycle (millions of years) is important only in controlling long-term variations in levels of atmospheric carbon dioxide.

The Industrial Revolution marked a turning point in the balance of energy in the earth's climate system. The rise of industry and technology resulted in an increase in the burning of wood and coal that serve as *sources* of carbon into the atmosphere. At the same time, changes in land use, including deforestation, worsened this imbalance by removing a potential sink for carbon. Carbon sources and sinks refer to parts of the carbon cycle that transfer carbon into and out of the atmosphere, respectively. Today, human activity is responsible for releasing approximately 9 billion metric tons of carbon per year into the atmosphere, with the ocean and land absorbing approximately half of that carbon (Le Quere et al. 2009). The accumulation of atmospheric carbon is analogous to a rising pool of water in a bathtub with a drain that is partially closed. The drain represents the sinks for atmospheric carbon in the oceans and the land, whereas the output of the faucet represents carbon in the atmosphere from sources such as fossil fuel burning and land-use change. If water comes out of the faucet faster than the drain can remove it, the water level in the bathtub will rise. Similarly, if sources of carbon outweigh the sinks, the amount of carbon in the atmosphere will continue to rise.

Electricity generated by the burning of fossil fuels accounted for more than a third of manmade carbon dioxide released in 2011 in the United States (EPA 2013b). In contrast, other energy sources—including nuclear, solar, wind, hydroelectric, and geothermal energy sources—emit minimal, if any, greenhouse gases. Using biomass to generate electricity can also reduce net

Box 2.2
Charles David Keeling, 1928–2005

In 1958, Charles David Keeling, a professor at the Scripps Institution of Oceanography at the University of California at San Diego, began to collect a continuous record of atmospheric carbon dioxide concentrations from towers on the Mauna Loa Observatory in Hawaii. Prior to his observations, it was unclear whether anthropogenic CO_2 emissions accumulate in the atmosphere or are absorbed by vegetation and the oceans. Keeling's measurements have confirmed increased CO_2 levels and are a leading piece of evidence of anthropogenic effects on atmospheric chemistry. Many consider his work to be the single most important environmental dataset of the twentieth century.

Box 2.3
The Colonials

Colonial Americans were sensitive to British criticism of the cold American climate and, according to science historian James Fleming (1998), argued that their climate was improving as the forests were cleared. In 1721, Cotton Mather believed that the North American region was getting warmer: "Our cold is much moderated since the opening and clearing of our woods, and the winds do not blow roughly as in the days of our fathers, when water, cast up into the air, would commonly be turned into ice before it came to the ground." Benjamin Franklin agreed: "cleared land absorbs more heat and melts snow quicker," but concluded that many years of observations would be necessary to settle the issue of climatic change (Fleming 1998, 24).

greenhouse gas emissions. Forest trimmings, orchard prunings, livestock manure, cheese-processing waste, restaurant waste, gases from wastewater treatment plants, and gas from landfills are all forms of biomass. Biomass generally removes carbon dioxide from the air when growing and emits methane, which is a more potent greenhouse gas than carbon dioxide, when it decays. When managed sustainably, displacing the use of fossil fuels with full combusted or anaerobically digested biomass to generate electricity can reduce net greenhouse gas emissions.

The transportation sector is the second biggest source of carbon dioxide emissions in the United States. Every gallon of gasoline consumed releases about 2.5 kilograms (5.5 pounds) of carbon into the atmosphere. Fuel economies of many automobiles have improved dramatically over the past few decades because of technological improvements, but the carbon dioxide emitted from vehicles in the United States is still approximately

Box 2.4
Guy Stewart Callendar, 1898–1964

Beginning in 1938, the British engineer and scientist Guy Stewart Callendar identified important links between the burning of fossil fuels and global warming (Fleming 2007). He compiled weather data from stations around the world that showed a warming trend of 0.5°C (0.9°F) in the early decades of the twentieth century. He estimated a 10 percent increase in atmospheric CO_2 levels based closely on the amount of fuel burned between 1900 and 1935, and hypothesized the warming was due to the increase in atmospheric CO_2 concentrations. Today, the theory is called the *Callendar effect*.

equal to that emitted from all sources in India (even though the population of India is nearly four times that of the United States).

In addition to CO_2 emissions, more than half of today's methane emissions are attributable to human activities, such as burning biomass, cultivating rice, creating landfills, and managing livestock (Solomon et al. 2007, chapter 7.4.1). Although much lower in atmospheric concentrations compared to CO_2, methane is the second most important greenhouse gas due to its strong GWP. Atmospheric concentrations of methane have increased by more than 150 percent since the beginning of the Industrial Revolution as a direct result of human activities (EPA 2013a, Introduction, 5). Methane releases from agricultural and natural sources are expected to increase in a warming planet, regardless of human activity. Studies argue that warming may melt huge expanses of high-latitude permafrost and release large reserves of methane that have been locked away in deep soils (Zimov, Schuur, and Chapin 2006) and beneath the oceans (Pearce 2007).

Aerosols

In addition to gases, the atmosphere contains suspended solid and liquid particles called *aerosols*. Aerosols can range in size from tiny molecular clusters to particles visible to the human eye. The principal sources of aerosols include both natural sources, such as dust, vegetation, sea spray, and volcanoes, as well as anthropogenic sources from fossil-fuel combustion and biomass burning. Like greenhouse gases, aerosols play a role in the earth's climate and have increased in the atmosphere since the Industrial Revolution. However, unlike greenhouse gases that act primarily on *longwave* radiation and have long atmospheric lifetimes, aerosols primarily interact with *shortwave* radiation and have short lifetimes.

Unlike greenhouse gases that are not visible to the human eye, aerosols can be seen in sunlight and often play a role in brilliant autumn-hued sunsets. Aerosols come in different colors and differ in their ability to reflect and absorb sunlight. Light-colored aerosols reflect incoming solar radiation, thereby decreasing the amount of energy that reaches the earth's surface. Much as wearing light-colored clothing keeps you cooler when you're in the sun, these aerosols tend to cool the planet. In contrast, dark aerosols that are a byproduct of incomplete fossil fuel and biomass combustion—such as black carbon—absorb solar radiation and heat the atmosphere. Additionally, when black carbon from the atmosphere is deposited on mountain snowpack and glaciers, it increases the amount of solar radiation absorbed and contributes to additional warming. Recent studies show that the additional energy absorbed by black carbon is nearly half the total warming influence of anthropogenic greenhouse gases, second only to carbon dioxide in terms of its contribution to global warming (Ramanathan and Carmichael 2008; Bond et al. 2013).

Over the last two hundred years sulfate aerosols, a byproduct of emissions from fossil-fuel combustion, have increased enormously. In parts of the world that consume large amounts of sulfur-rich coal, such as China and central Europe, sulfate emissions partially offset the warming effects of high levels of greenhouse gases in the short term. But emitting these aerosols to lessen the enhanced greenhouse effect, as some geoengineers and others suggest, is problematic for several reasons. First, sulfates combine with water vapor to form sulfuric acid, the principle component in acid rain. Second, the short lifetimes of aerosols (days to weeks) result in regional impacts, while long-lived greenhouse gases are diffused across the globe. Third, efforts to counter greenhouse gas–fueled warming with aerosols would require massive, continual, and widespread emissions. Finally, loading our atmosphere with aerosols may result in undesirable and unforeseen consequences on a host of processes crucial to life on our planet. The concept of adding aerosols to the atmosphere to lessen the warming of the planet is further discussed in the section on geoengineering in chapter 5 of this book.

In contrast, due to the strong warming influence of black carbon and the fact that aerosol lifetimes are very short, some have suggested that efforts to curb black carbon emissions are a cost-effective route to temper warming trends. Emissions of black carbon are particularly high in southern Asia, where 40 percent of black carbon emissions come from solid biomass used for residential cooking (EPA 2012). Efforts to provide alternatives to such fuels in parts of the developing world may not only have immediate benefits to climate, but also may reduce the health impacts associated with poor air quality in such regions. In comparison to the worldwide effort to reduce carbon emissions, many believe that reduction of black carbon emissions is a more acceptable and readily available option for mitigating climate change.

Climate Change as an Environmental Problem

Climate change was recognized as a global environmental problem for the first time in the late 1970s. James Hansen, an American climatologist, advanced the Callendar hypothesis that fossil fuel burning was heating the planet. At the first World Conference on Climate, held in Geneva in 1979, many scientists warned that climate changes driven by human activity would detrimentally affect humankind and the environment. The conference attendees invited all heads of state to heed their warnings about climate change and enact "necessary policies for the well-being of humanity." The worldwide response to this invitation and later warnings are discussed in chapter 5.

The Climate Puzzle

Numerous natural and anthropogenic factors have produced today's climate. The studies of past climates (in a field called *paleoclimatology*) and modern climate (called *climatology*) describe the processes and patterns of the earth's climate system and the forces that drive climate variability and change. If we can thoroughly understand past variations, we are more apt to understand recently observed and projected changes in climate. Natural variations in climate—including oscillations in large-scale wind patterns, changes in oceanic circulation, solar variability, and variations in the earth's orbit—occur on timescales ranging from years to decades to tens of thousands of years.

Box 2.5
James E. Hansen

Although James Hansen's early research focused on the atmosphere of Venus, he soon was drawn to the exciting and innovative research being done regarding earth's climate. He has since added greatly to our body of knowledge on climate change through the development of climate models that numerically simulate global climate and that help us understand the complexities of the climate system and test hypotheses of climate change. Some of Hansen's initial projections of global warming trends made in the early 1980s have been fairly accurate and have been validated by the continued warming of the planet. Based on his research, Hansen concludes:

> Climate system inertia means that it will take several centuries for the eventual extreme global warming mentioned above to occur, if we are so foolish as to burn all of the fossil fuel resources. Unfortunately, despite the ocean's thermal inertia, the transient climate phase this century, if we continue business-as-usual fossil fuel burning, is likely to cause an extended phase of extreme climate chaos. As ice sheets begin to shed ice more and more rapidly, our climate simulations indicate that a point will be reached when the high-latitude ocean surface cools while low-latitudes surfaces are warming. An increased temperature gradient, i.e., larger temperature contrast between low and high latitudes, will drive more powerful storms. ... The science of climate change, especially because of the unprecedented human-made climate forcing, includes many complex aspects. This complexity conspires with the nature of reporting and the scientific method itself, with its inherent emphasis of caveats and continual reassessment of conclusions, to make communications with the public difficult, even when the overall picture is reasonably clear. (2013)

Hansen for many years was the director of the NASA Goddard Institute for Space Studies in New York. In 2013 he announced that he was retiring from NASA to devote himself to becoming an activist for climate change.

Natural Climate Variability

Observations suggest that global climate change is currently underway. Paleoclimatic data reveal that change has occurred in the global climate throughout the earth's history. Modern climate observations represent a mere fraction of climate history. How can we know whether the current changes are part of the natural ebb and flow of global climatic changes, or if they are extremely unusual against the backdrop of the earth's history, and thus point to human-caused factors? Like a detective trying to solve a case, paleoclimatologists search for pieces of evidence to solve the climate puzzle. Although the thermometer wasn't invented until the seventeenth century, scientists are able to reconstruct climate before the era of modern instruments through *proxy data*. Proxy data are essentially "natural" historical recording systems of climatic conditions and the chemical makeup of the atmosphere.

Proxy data are found in objects familiar to most of us, such as trees and fossil records, and objects not as familiar to us, such as lake and ocean sediments. The most widely recognizable source of proxy data is tree rings, which record year-to-year variations in tree growth. Some trees are sensitive to energy requirements (temperature), while others might be sensitive to moisture requirements (precipitation). For example, trees in a semi-arid climate might be able to grow at a faster rate during unusually wet years, whereas those near the upper tree line, where moisture is usually not a limiting factor, might grow at a faster rate during warmer periods. Trees can provide a trace of climate over their lifespan of some hundreds to even thousands of years in some cases. Other sources of proxy data include sediments, ice cores, corals, and the geologic record. As mentioned previously, cores extracted from glaciers

Box 2.6
Svante Arrhenius, 1859–1927

"Is the mean temperature of the ground in any way influenced by the presence of the heat-absorbing gases in the atmosphere?" (Arrhenius 1896, 237). In 1895, the Swedish chemist Svante Arrhenius presented an answer to this question to the Stockholm Physical Society (Fleming 1998). Arrhenius did not pursue an answer out of any great concern for increasing levels of CO_2; rather, he was attempting to explain temperature changes between glacial and interglacial periods.

Arrhenius formulated a heat budget for the planet in which changes in the atmospheric levels of carbon dioxide are matched by changes in surface temperature. With this model, he concluded that the temperature of the Arctic region would increase by 8–9°C (14°–16°F) if atmospheric CO_2 concentrations tripled from preindustrial levels. Arrhenius (1908, 51) went on to describe the "hot-house theory" (greenhouse effect), calling on earlier studies suggesting that the earth's surface temperature would be about 33°C (59°F) cooler in the absence of atmospheric gases. In his work he also noted that the increased production of CO_2 "by the advances of industry" could alter climate "to a noticeable degree in the course of a few centuries" (1908, 54). However, he maintained that this increased use of fossil fuels could be advantageous for the climate, resulting in "ages with equable and better climates, especially as regards the colder regions of the Earth" (1908, 63).

In 1899, Nils Ekholm, an associate of Arrhenius, suggested that rates of coal burning could double the concentration of atmospheric CO_2 and "undoubtedly cause a very obvious rise of the mean temperature of the Earth." In Arrhenius's 1908 book, *Worlds in the Making*, he popularized this hypothesis. Arrhenius considered that such emissions would be fortuitous given that "a new ice period ... will drive us from our temperate countries into the hotter climates of Africa." Arrhenius speculated on a "virtuous circle" in which fossil fuel burning could delay a rapid return to the ice ages. Arrhenius's CO_2 theory of climate change fell out of scientific favor and was not revived in its modern form until the mid-1950s (Fleming 1998).

can extend hundreds of meters below ground and reveal the chemical composition of the air bubbles imprisoned in glacial ice. Scientists can use pollen samples from lakebeds to identify vegetation that flourished nearby millions of years ago, and use this to infer climatic conditions suited to such vegetation.

Climate reconstructions have greatly advanced our understanding of the climate system and natural climate cycles. Slow changes in earth's orbit are the leading initiator of the glacial-interglacial cycles that have been the pacemaker of global climate over the past two million years; each cycle requires approximately 100,000 years. The glaciers covering North America and Europe during the last glacial period began receding 18,000 years ago.

The transition between glacial conditions and the interglacial conditions that we are experiencing today was not a smooth one; it was marked instead by a number of abrupt changes that occurred over decades (Severinghaus et al. 1998), though not as quickly as those depicted in the 2004 movie *The Day after Tomorrow*.

Some models have suggested that an abrupt change in climate could slow down deep-ocean circulation (called *thermohaline circulation*). This circulation is driven by slight differences in *water density* that arise in response to variations in water temperature and *salinity*, or salt content. At the surface of the ocean, winds guide oceanic currents that work with the atmosphere to redistribute heat across the globe. For example, the Gulf Stream pulls warm salty water away from low latitudes and toward the high latitudes, releasing heat into the North Atlantic and moderating climate across northwestern Europe. As this salty water moves into higher latitudes, it cools and increases in density and ultimately sinks to the bottom of the ocean, thereby churning the thermohaline circulation.

Oceanic circulation is more apt to be maintained if changes are of a gradual nature. By contrast, discharge of freshwater into the North Atlantic that could result from rapid melting of freshwater glacial ice may shut down circulation by decreasing the density of surface water (McCarthy et al. 2001, 17).

Thirteen thousand years ago, global temperatures were rebounding after the previous glacial period. Suddenly, however, within a decade, glacial conditions rapidly returned to the high latitudes of North America and Europe as high northern latitudes cooled nearly 6°C (11°F). This return to glacial conditions is known as the Younger Dryas, named for a flower tolerant of cold conditions found in the pollen record across much of Europe during the period. Scientists hypothesize that this abrupt cooling resulted from an influx of fresh water into the salty waters of the North Atlantic released after the melting of the Laurentide ice sheets over eastern Canada. This freshening of the North Atlantic slowed down thermohaline circulation and the Gulf Stream that delivers warm water from the balmy subtropics to northwestern Europe. Archaeologists note that the onset of this cooler and drier period coincided with the beginning of agriculture in northern Mesopotamia (Calvin 2002). Fluctuations in the climate likely had an adverse impact on the food supply of hunter-gatherers, creating an incentive for developing agriculture as a more stable and reliable food source.

The Medieval Warm Period and the Little Ice Age are the two most noted "natural" climate cycles of the last millennia. These variations in climate were modest, compared to the Younger Dryas, yet had impacts on growing population centers in Europe. During the Medieval Warm Period (in the tenth to fourteenth centuries), warmer Northern Hemisphere temperature (similar to temperatures during the twentieth century, but cooler than present) allowed for the Viking colonization of Greenland. However, the onset of the Little Ice Age (in the fifteenth to nineteenth centuries) ushered in cooler temperatures, which resulted in the collapse of these colonies. Northern Hemisphere temperatures were about 1°C (1.8°F) cooler than twentieth-century conditions, and were most notable across parts of Europe. These seemingly benign changes in global climate posed an additional stressor to society, impacting agricultural productivity and leading to starvation and an overall deterioration in human health. The Medieval Warm Period and Little Ice Age are hypothesized to be a result of a combination of solar, volcanic, and oceanic variability.

On much shorter timescales, phenomena such as the El Niño Southern Oscillation and volcanic eruptions have important but short-lived impacts on global climate. El Niño's Southern Oscillation is a coupled atmosphere-ocean phenomenon that channels interannual (year-to-year) fluctuations in ocean temperatures over the tropical East Pacific into global fluctuations in climate. For example, under El Niño conditions, when ocean surface temperatures off the coast of Peru are unusually warm during winter, the southern half of the United States experiences cooler and wetter winters than normal, while the northern half of the country from the Great Lakes westward to the Pacific Northwest experiences warmer than normal winters due to a change in the location of the storm track. Typically, El Niño conditions result in a spike in mean annual global temperatures by a few tenths of a degree, while La Niña conditions (the reverse of El Niño conditions) result in a short-lived cooling of global temperatures. Year-to-year variations in the natural climate system are one reason why increased greenhouse gas concentrations do not result in yearly increases, though long-term warming trends are clear.

Volcanic eruptions have long been implicated in climate variations. Volcanic eruptions release huge quantities of ash, aerosol droplets, and volcanic gas, including sulfur dioxide and carbon dioxide, into the atmosphere. The sulfur dioxide gas masses can be injected high into the atmosphere, where they form bright reflective aerosols that shield the earth from the sun's rays and lead to a short-lived (up to a few years) global cooling (USGS 2012). This effect was noted in the last few decades following the eruption of El Chichon in Mexico in 1982 and Mount Pinatubo in the Philippines in 1991. These recent volcanic eruptions pale in comparison to the volcanic activity from Mount Tambora that preceded the so-called Year without a Summer in 1816. That summer was marked by unusually cool conditions across much of Europe and the eastern United States. Cold temperatures that year resulted in widespread crop failure across developing North American settlements and exacerbated food crises and epidemics across much of Europe; it has been labeled "the last great subsistence crisis in the Western world" (Post 1977, i).

Global Climate Models

Scientists who study changes in the earth's climate system cannot follow classical experimental methods that include a "control group" because we only have one earth. In the absence of a control group, scientists study dynamics of global climate change by constructing numerical computer models of the climate system. Vast improvements in our understanding of the climate system have been made over the last three decades through a combination of advancements in computation capabilities, high-quality observations, theory, and interdisciplinary science.

Global climate models, or GCMs, are physics-based numerical models that represent our best understanding of the processes and interactions of components of the earth's climate system. These models have been used to explain past changes in climate, and are used today in seasonal climate outlooks and long-term climate predictions. Several modeling groups across the world have developed GCMs. While this may appear to be a redundant exercise, these GCMs have varying levels of complexity and computation demand, and thus yield slightly different results when the same experiment is conducted. Using several models to conduct the same experiment is similar to how biologists use many lab rats in an experiment to examine how well a new serum works to cure an illness. This "probabilistic approach" is generally preferred over relying on the outcome of a single model. It not only tests the robustness of outcomes but also can better frame the odds of varying levels of change.

The fundamental issue in climate change is whether observed changes in climate have resulted from natural or human forcing, or from some combination of the two. Numerous studies have been conducted to determine the extent to which climate changes realized during the twentieth century are the result of natural variability (Karl and Trenberth 2003). Models that account solely for natural forcing, including volcanic eruptions and solar variability, are unable to replicate observed twentieth-century changes in temperature, ocean heat content, and global processes. Models are only able to replicate the observed changes in climate after including anthropogenic forcing, including the observed increases in greenhouse gases and aerosol concentrations—providing strong evidence that a majority of the warming observed during the twentieth century is attributable to increased levels of atmospheric greenhouse gases. This evidence was presented to the White

House in spring 2001 in a report that concluded, "Greenhouse gases are accumulating in earth's atmosphere as a result of human activities, causing surface air temperatures and subsurface ocean temperatures to rise. ... The changes observed over the last several decades are likely mostly due to human activities" (National Academy of Sciences Committee 2001, 1). In 2007, the Intergovernmental Panel on Climate Change (IPCC) stated, "Since the start of the industrial era (about 1750), the overall effect of human activities on climate has been a warming influence. The human impact on climate during this era greatly exceeds that due to known changes in natural processes, such as solar changes and volcanic eruptions" (Solomon et al. 2007, FAQ 2.1).

Scientists have used many sources of data to investigate the dynamics of the climate system in order to expand and improve the scientific observations that are the basis for GCM calculations. In the absence of a dynamic climate system, it would be quite easy to calculate the net change in global temperature (a 1°C warming) due to a doubling of CO_2. Unfortunately, the climate system is much more complex. A slight warming in surface temperature results in a cascade of processes known as *climate feedbacks* that interactively alter components of the system and can affect the original change. For example, in a warming planet, the increase in evaporation and the ability of a warmer atmosphere to hold more water vapor both strengthen the greenhouse effect, thereby amplifying warming. However, increased water vapor may also lead to increased cloud formation, reflecting solar radiation and moderating the warming. Scientists suggest that clouds are the largest wildcard in climate due to their ability to trap outgoing thermal radiation (and thus help to warm the planet) and reflect incoming solar radiation (thus helping to cool the planet; Emanuel 2012). Other uncertainties include the carbon cycle and biotic response to climate change.[1]

Existing models do not currently account for the release of large quantities of methane, a gas with high heat-trapping potential, from the immense tracts of permafrost now melting due to global warming (Zimov et al. 2006; Williamson, Saros, and Schindler 2009; Ruppel 2011). Additionally, a recent study reveals that large urban areas produce enough heat during winter months to influence the width of the jet stream and the nature of other climate processes (Scripps Institution of Oceanography 2013). Were all these elements taken into consideration in the development of climate change models, the projected heat increases and the potential consequences of global warming would be even worse than current models indicate.

Climate feedbacks play a critical role in future climate scenarios by indicating how sensitive climate is to changes in the amount of energy to the system. Our current best estimate of the earth's *climate sensitivity*, a measure of the change in global temperature in response to a doubling of atmospheric CO_2, is an increase in temperature between 2–4.5°C (3.6–8.1°F; Alley et al. 2007). This increase is substantially larger than the 1°C (1.8°F) warming in global mean temperature projected to occur due only to a doubling of CO_2.

Finally, future climate change depends on the trajectory of global demographics as well as energy sources, land use, and the effects of globalization. The IPCC's emission scenarios (IPCC 2000) were used to form a basis for projected changes in emissions of greenhouse gases and aerosols. Scenarios range from optimistic future conditions, where population levels stabilize and levels of greenhouse gas emissions are reduced, to pessimistic future conditions in which population and greenhouse gas emissions continue to grow. More recently, an equivalent set of trajectories called *representative concentration pathways* was devised to provide a more flexible set of scenarios. These representative concentration pathways focus on the amount of additional energy trapped by the earth-atmosphere system. These scenarios are run through a set of GCMs to provide scientists with a confidence range for projections of future climate change.

The IPCC produces reports, based on hundreds of scientific studies, that summarize the current understanding of observed climate change. The reports provide climate projections and describe climate impacts and adaptation and mitigation measures. The fourth assessment report, released in 2007, stated, "Warming of the climate system is unequivocal, as is now evident from observations of increases in global average air and ocean temperatures, widespread melting of snow and ice and rising global average sea level" (IPCC 2007a, 2).[2] According to the National Oceanic and Atmospheric Administration (2013), observations show that "the combined average temperature over global land and ocean surfaces for May 2013 tied with 1998 and 2005 as the third warmest on record, at 0.66°C (1.19°F) above the 20th century average of 14.8°C (58.6°F)." Furthermore, observed global temperature trajectories have been broadly consistent with the GCM-based projections established by James Hansen in the 1980s. Perhaps more alarming is the fact that the changes have occurred faster than predicted. For example, observed increases in sea ice retreat, snow cover loss, and mass loss of Greenland's ice sheet have exceeded climate projections to date (e.g., Stroeve et al. 2007, Pierce et al. 2008). Scientists have revised the IPCC's predictions for Arctic sea ice (IPCC 2007b, 659), and now predict the Arctic may be nearly ice-free in summer between 2030 and 2050, decades earlier than originally predicted (Wang and Overland 2009; IPCC 2013, 23).

Conclusion

We have reviewed the basics of climate science, emphasizing how the earth's climate system balances both energy and the atmospheric composition to sustain life on the planet. This balance is threatened by an abrupt rise in greenhouse gas concentrations since the Industrial Revolution. Ambient levels of carbon dioxide increased from about 280 ppm in the mid-1800s to approximately 400 ppm in 2013. Levels of atmospheric CO_2 are anticipated to reach between 600 and 1,000 ppm by the end of the twenty-first century.

The science of climate change can be thought of as a movie that has been made by hundreds of directors and that takes viewers from billions of years ago to the present. But it is a film with many blurry images and empty frames. The goal of research in climate science is to refine these images and fill in the missing frames to provide context to the plot. As the movie chronicles actions that have brought us to the current era of global climate change, the challenge is to decipher the trajectory of the story in time to avoid a disastrous ending. The great majority of experts in the field believe that current evidence definitively identifies humankind as the story's antagonist. The movie, of course, is left unfinished and has alternate endings. It leaves viewers wondering how the story will end.

Anticipating the outcome of the story is difficult. Prediction of the future is constrained by the limitations of scientific knowledge, the complexity of the climate system, uncertain population growth, an interconnected global economy, advances in technology, and changes in politics and policies.

Given these uncertainties, conclusions that scientists reach on future changes in climate are expressed in terms of probabilities and risks.

How we should proceed? Can climate projections inform us regarding what actions, if any, should be taken to reduce risks and impacts of climate change? How do we weigh the costs of action against the costs of inaction? Potential routes to tackle this dilemma are addressed in subsequent chapters.

Notes

1. Another difficulty in the creation of accurate climate models is the uncertainty inherent in the Earth's climate. Emanuel explains: "The amount of uncertainty in such [climate] projections can be estimated to some extent by comparing forecasts made by many different models" (Emanuel 2012, 46). He adds, "This exercise has been repeated using many different climate models, with the same qualitative result: one cannot accurately simulate the evolution of the climate over the last 30 years without accounting for the human input of sulfate aerosols and greenhouse gases. This is one (but by no means the only) important reason that almost all climate scientists today believe that man's influence on climate has emerged from the backgrounds of natural variability" (Emanuel 2012, 48–49).

2. Similarly, the fifth assessment report, released in 2013, stated, "Warming of the climate system is unequivocal, and since the 1950s, many of the observed changes are unprecedented over decades to millennia. The atmosphere and ocean have warmed, the amounts of snow and ice have diminished, sea level has risen, and the concentrations of greenhouse gases have increased" (IPCC 2013, 2).

References

Alley, Richard, Terje Berntsen, Nathaniel L. Bindoff, Zhenlin Chen, Amnat Chidthaisong, Pierre Friedlingstein, Jonathan M. Gregory, et al. 2007. IPCC 2007: Summary for Policymakers. In *Climate Change 2007: The Physical Science Basis. Contribution of Working Group I* to the Fourth Assessment Report of the Intergovernmental Panel on Climate Change, edited by S. Solomon, D. Qin, M. Manning, Z. Chen, M. Marquis, K. B. Averyt, M. Tignor, and H. L. Miller. New York: Cambridge University Press.

Arrhenius, Svante. 1896. On the influence of carbonic acid in the air upon the temperature of the ground. *London, Edinburgh, and Dublin Philosophical Magazine and Journal of Science* (5th ser. April) 41: 237–275.

Arrhenius, Svante. 1908. *Worlds in the Making: Evolution of the Universe.* Translated by H. Borns. New York: Harper.

Bond, T. C., S. J. Doherty, D. W. Fahey, P. M. Forster, T. Berntsen, B. J. DeAngelo, M. G. Flanner, et al. 2013. Bounding the role of black carbon in the climate system: A scientific assessment. *Journal of Geophysical Research* 18:5380–5552. doi:10.1002/jgrd.50171.

Calvin, William H. 2002. *A Brain for all Seasons: Human Evolution and Abrupt Climate Change.* Chicago: University of Chicago Press.

Emanuel, Kerry. 2012. *What We Know About Climate Change.* 2nd ed. Boston, MA: MIT Press.

Environmental Protection Agency. 2012. Reducing Black Carbon Emissions in South Asia: Low Cost Opportunities. Office of International and Tribal Affairs. http://www.unep.org/ccac/Portals/24183/docs/BlackCarbonSAsiaFinalReport5.22.12.pdf, accessed July 8, 2013.

Environmental Protection Agency. 2013a. Inventory of US Greenhouse Gas Emissions and Sinks 1990–2011. EPA 430-R-13–001. April 12.

Environmental Protection Agency. 2013b. Overview of Greenhouse Gases Carbon Dioxide Emissions. http://www.epa.gov/climatechange/ghgemissions/gases/co2.html. Last update: Monday September 09, 2013, accessed September 16, 2013.

European Space Agency Ozone Climate Change Initiative. 2013. *Ozone_cci Newsletter* 4 (February):2–3. http://www.esa-ozone-cci.org, accessed July 8, 2013.

Fleming, James Rodger. 1998. *Historical Perspectives on Climate Change.* New York: Oxford University Press.

Fleming, James Rodger. 2007. *The Callendar Effect: The Life and Work of Guy Stewart Callendar.* Boston: American Meteorological Society.

Hansen, James. 2013. Making things clearer: Exaggeration, jumping the gun, and the Venus syndrome: Essay on climate science communication. Recent Communications. April 15. http://www.columbia .edu/~jeh1/, accessed July 8, 2013.

Heinlein, Robert. 1978. *The Notebooks of Lazarus Long*. New York: Spectrum Literary Agency.

Intergovernmental Panel on Climate Change. 2000. *Special Report. Emissions Scenarios: Summary for Policy Makers.* A Special Report of IPCC Working Group III. http://www.ipcc.ch/pdf/special-reports/spm/ sres-en.pdf, accessed July 8, 2013.

Intergovernmental Panel on Climate Change. 2007a. *Climate Change 2007: Synthesis Report. Summary for Policy Makers.* Contribution of Working Groups I, II and III to the Fourth Assessment Report of the Intergovernmental Panel on Climate Change [Core Writing Team, Pachauri, R.K and Reisinger, A. (eds.)]. http://www.ipcc.ch/pdf/assess-ment-report/ar4/syr/ar4_syr_spm.pdf, accessed September 11, 2013.

Intergovernmental Panel on Climate Change. 2007b. *Climate Change 2007: Impacts, Adaptation and Vulnerability. Chapter 15: Polar Regions (Arctic and Antarctic).* http://www.ipcc.ch/pdf/assessment-re-port/ar4/wg2/ar4-wg2-chapter15.pdf, accessed September 16, 2013.

Intergovernmental Panel on Climate Change. 2013. Working Group I Contribution to the IPCC Fifth Assessment Report Climate Change 2013: The Physical Science Basis Summary for Policymakers. http:// www.climatechange2013.org/spm, accessed December 16, 2013.

Karl, Thomas R., and Kevin E. Trenberth. 2003. Modern global climate change. *Science* 302 (December 5):1719–1723.

Lazo, J. K., R. E. Morss, and J. L. Demuth. 2009. 300 billion served: Sources, perceptions, uses, and values of weather forecasts. *Bulletin of the American Meteorological Society* 90:785–798.

Le Quere, C., and Associates. 2009. Trends in the sources and sinks of carbon dioxide. *Nature Geoscience* 2 (November):831–836.

Lüthi, D., M. Le Floch, B. Bereiter, T. Blunier, J.-M. Barnola, U. Siegenthaler, D. Raynaud, et al. 2008. High-resolution carbon dioxide concentration record 650,000–800,000 years before present. *Nature* 453:379–382.

McCarthy, James J., Osvaldo F. Canziani, Neil A. Leary, David J. Dokken, and Kasey S. White, eds. 2001. *Climate Change 2001: Impacts,*

Adaptation and Vulnerability. Cambridge, UK: Cambridge University Press.

Merall, Zeeya. 2013. Quantum gas goes below absolute zero. Nature News. January 3. http://www.nature.com/news/quantum-gas-goes -below-absolute-zero-1.12146, accessed July 6, 2013.

Mohan, Geoffrey. 2013. Carbon dioxide levels in atmosphere pass 400 milestone, again. *Los Angeles Times*. May 20. http://articles .latimes.com/print/2013/may/20/science/la-sci-sn-carbon-dioxide -400-20130520, accessed September 16, 2013.

National Academy of Sciences. 2010. *Advancing the Science of Climate Change*. By America's Climate Choices: Panel on Advancing the Science of Climate Change; National Research Council. Washington, DC: National Academies Press.

National Academy of Sciences Committee on the Science of Climate Change. 2001. *Climate Change Science: An Analysis of Some Key Questions*. Washington, DC: National Academies Press.

National Oceanic and Atmospheric Administration. 2012. The NOAA Annual Greenhouse Gas Index. Updated summer 2012.

National Oceanic and Atmospheric Administration. 2013. State of the Climate: Global Analysis for May 2013. National Climatic Data Center (June). http://www.ncdc.noaa.gov/sotc/global/2013/05/, accessed July 8, 2013.

Pearce, Fred. 2007. *With Speed and Violence: Why Scientists Fear Tipping Points in Climate Change*. Boston: Beacon Press.

Pierce, D. W., Tim P. Barnett, Hugo G. Hidalgo, Tapash Das, Celine Bonfils, Benjamin D. Santer, Govindasamy Bala, et al. 2008. Attribution of declining western U.S. snowpack to human effects. *Journal of Climate* 21 (December):6425–6444.

Post, John Dexter. 1977. *The Last Great Subsistence Crisis in the Western World*. Baltimore, MD: Johns Hopkins University Press.

Ramanathan, V., and G. Carmichael. 2008. Global and regional climate changes due to black carbon. *Nature Geoscience* 1 (March):221–227.

Ruppel, Carolyn D. 2011. Methane hydrates and contemporary climate change. *Nature Education Knowledge* 3 (10):29.

Scripps Institution of Oceanography. 2013. Urban heat has large-scale climate effects. January 30. http://explorations.ucsd.edu/research-high

lights/2013/urban-heat-has-large-scale-climate-effects/, accessed July 8, 2013.

Seager, R. 2006. The source of Europe's mild climate. *American Scientist* 94 (July–August):334–341.

Severinghaus, J. P., T. Sowers, E. J. Brook, R. B. Alley, and M. L. Bender. 1998. Timing of abrupt climate change at the end of the Younger Dryas interval from thermally fractionated gases in polar ice. *Nature* 391 (January):141–146.

Solomon, S., D. Qin, M. Manning, Z. Chen, M. Marquis, K. B. Averyt, M. Tignor, et al., eds. 2007. *Climate Change 2007: The Physical Science Basis. Contribution of Working Group I to the Fourth Assessment Report of the Intergovernmental Panel on Climate Change.* Chapter 7.4.1, Methane. New York: Cambridge University Press.

Stroeve, J., M. M. Holland, W. Meier, T. Scambos, and M. Serreze. 2007. Arctic sea ice decline: Faster than forecast. *Geophysical Research Letters* 34 (May):L09501. doi:10.1029/2007GL029703.

Tans, Pieter, and Ralph Keeling. 2013. Trends in atmospheric carbon dioxide. National Oceanic and Atmospheric Administration Earth System Research Laboratory (August). http://www.esrl.noaa.gov/gmd/ccgg/trends/, accessed September 16, 2013.

US Geological Survey. 2012. Volcanic gases and climate change overview. Volcano hazards program. Last update January 27. http://volcanoes.usgs.gov/hazards/gas/climate.php, accessed July 6, 2013.

Wang, M., and J. E. Overland. 2009. A sea ice free summer Arctic within 30 years? *Geophysical Research Letters* 36:L07502. doi:10.1029/2009GL037820.

Williamson, Craig E., Jasmine E. Saros, and David W. Schindler. 2009. Sentinels of Change. *Science* 323:887–888.

World Meteorological Organization. 2002. *Executive Summary: Scientific Assessment of Ozone Depletion. Global Ozone Research and Monitoring Project Report No. 47.*

Zimov, Sergey A., Edward A. G. Schuur, and F. Stuart Chapin, III. 2006. Permafrost and the global carbon budget. *Science* 312 (5780):1612–1613.

3

Climate-Change Effects, Adaptation, and Mitigation

John T. Abatzoglou, Crystal A. Kolden, Joseph F. C. DiMento, Pamela Doughman, and Stefano Nespor

Modeling of climate provides information about the physical manifestation of climate change at many levels, from the global to the local. In order to understand these changes, additional steps are needed to translate physical changes into climate effects and to assess risk avoidance through mitigation and adaptation efforts.

Overall Global Consequences of Climate Change

The Intergovernmental Panel on Climate Change (IPCC) Fifth Assessment report released in 2013 provided strengthened support of widespread observed changes in global climate. Decadal global mean surface temperatures have progressively increased over the last three decades and were warmer than any decade since 1850 (IPCC 2013, SPM-3). The last month with global mean temperature below their twentieth-century average occurred in February 1985. Decreases in the mass of Greenland and Antarctic ice sheets, declines in sea-ice extent, and a contraction of Northern Hemisphere spring snow cover further substantiate a warming planet. The report further established clear links between human influences and many of the observed changes in climate.

These statements were released in spite of the relative lull of increases in global mean temperature over the past fifteen years. However, short-term records of global mean surface temperature are not a good proxy for assessing man-made changes in climate due to the natural year-to-year variability in climate. Dr. Richard Muller, who compiled an independent record of global temperature, described the fallacies of short-term records of climate change with the analog: "When walking up stairs in a tall building, it is a mistake to interpret a landing as the end of the climb" (2013). Indeed, the observational record of global mean temperature shows numerous perceived short-term "pauses" in warming that followed a continuation of long-term warming. With more warming already "in the pipeline" and atmospheric carbon dioxide concentrations escalating, the reality of a warming world and the consequences associated with it are increasingly apparent.

Global atmospheric carbon dioxide levels reached about 400 ppm in 2013. During 2012, CO_2 levels rose by 2.67 ppm over 2011 levels, according to the National Oceanic and Atmospheric Administration (NOAA). This increase represents the second-highest rise in CO_2 emission levels since record keeping began in 1959. Only 1998 showed a higher annual increase; during that year, CO_2 levels climbed by 2.93 ppm (Borenstein 2013). In its Fifth Assessment Report, the IPCC revealed that atmospheric CO_2 concentrations have increased by over 40 percent due to human activity since 1750, and by about 10 percent since 1990 (IPCC 2013, SPM-7).

In some scenarios, the IPCC (2007) estimated CO_2 concentrations may reach 600–1000 ppm by 2100. The IPCC estimated in 2007 that the ramp-up of atmospheric carbon dioxide could increase global average surface temperatures 1.8 to 4.0°C (3.2 to 7.2°F) by 2100 (Alley et al. 2007). In its Fifth

Assessment, the IPCC put the prospective range for the end of the twenty-first century as "*likely* to exceed 1.5 degrees C relative to 1850 to 1900" under a number of sets of assumptions (emphasis in original; IPCC 2013, SPM-15). This may not seem like a major change; after all, day-to-day changes in temperature at any given point on the globe regularly exceed 4°C (7.2°F). However, a 4°C (7.2°F) increase in global average surface temperature may be better understood by considering that current-day temperatures are only about 6°C (11°F) warmer on average than the temperatures during the most recent ice age 20,000 years ago. Such a rise would represent a significant change.

Climate impacts arise in response to significant departures from what we consider a *normal* range of variability. These deviations from *normal* include both extreme weather events—for example, a landfalling hurricane like Hurricane Katrina and its effects on lives and infrastructure in Louisiana and Mississippi, or Superstorm Sandy and its impacts along the East Coast—and extreme climate events like the drought across the midwestern United States in the summer of 2012 and its far-reaching influences on agriculture and commodities. Extreme events are the main channel by which climate and society interact, and they attract the most climate-related media reports. Data for global natural catastrophes show there is an upward-sloping trend in the number of extreme weather events each year, with the frequency of storms, flooding, and related events growing more quickly than the frequency of extreme temperature, drought, and forest fires (Munich Re Group 2011, 11). The United States alone experienced fourteen weather disasters in 2011 and eleven in 2012 that each totaled over $1 billion in damages, with disasters encompassing hurricanes, tornadoes, wildfires, flooding, and drought (National Climatic Data

Center 2013). However, the IPCC's Fifth Assessment Report indicates more uncertainty regarding the impact of climate change on hurricanes (Revkin 2013).

Extreme weather and climate shape society and ecosystems. For example, cities located in low-lying areas place restrictions on building in predefined flood plains, and infrastructure must be built to withstand a "one-hundred-year" flood. Ecosystems in California are drought-adapted, while those in northwestern Montana are cold-tolerant. The ability of society and ecosystems to adapt to climate change will be put to the test through extreme events. Extreme events are, by definition, low-risk high-impact events that are exceedingly rare in historic terms. Because climate change involves a shift in the statistics of weather, however, it is possible that even modest changes in climate will result in significant changes in the frequency and magnitude of extreme events (Meehl, Arblaster, and Tebaldi 2005; Diffenbaugh and Ashfaq 2010), which may be of greater importance than overall changes in the average climate. Containing the damage caused by climate change requires preparation for changes in the new normal and changes in the magnitude and frequency of extreme events.

What does a warmer world imply for human and ecological systems? This may be the most important question regarding climate change, as it may motivate mitigation and adaptation efforts. However, there is no straightforward answer. Effects can be detrimental and beneficial, direct and indirect. Impacts may occur through gradual changes in average climate as well as changes in extreme events. Regional and local changes in climate can be much more pronounced than globally averaged changes. To untangle some of these dynamics, the next section discusses expected consequences for human systems, followed by a discussion of expected consequences for natural systems.

Effects on Humans

The basic needs of human survival include food, water, and shelter. Climate enables the geographic existence of societies by providing resources to meet these needs. The ability of societies to sustain the essentials for a growing population during an era of changing climate is particularly challenged. Aside from climate change, natural variability in climate is an additional stressor in the network of factors (e.g., war, poverty, diseases) that affect the ability of populations to meet these needs. With climate change and a growing population, the challenge is more formidable, particularly for populations that struggle to meet the needs for survival and may not have resources to cope with an additional stressor.

Increased global population requires increased agricultural productivity. Despite increases in yield per acre as a result of mechanized agriculture, agricultural productivity is dependent on both available resources and a reliable climate. The projected impacts of climate change on agriculture are mixed. Increases in atmospheric carbon dioxide can act to stimulate crop productivity and increase water-use efficiency, albeit at the expense of nutritional quality. However, the direct effects of changes in precipitation and temperature are expected to alter agricultural productivity. Agriculture in areas with short growing seasons, such as Canada and Russia, where plants are currently growing below their optimum temperature, may benefit from a longer, warmer growing season. However, the 2010 drought and heat wave that resulted in a loss of over one-third of the Russian wheat crop hints there is a fine line between a longer growing season as a result of warming and potential heat and moisture stress that begets crop failure. Northward shifts in suitable climates are expected to push agriculture belts northward across the central United States, making the already marginal croplands of the US southern Great Plains unsustainable, and the prime cropland of the midwestern Great Plains less productive and profitable (e.g., Ortiz et al. 2008). Declining water availability in southern Asia, South America, and Africa will likely result in significant losses in major crops and threaten food sustainability (Lobell et al. 2008). Likewise, increases in drought frequency and magnitude will also threaten agriculture in southern Europe.

Climate scientists also expect that the *hydrologic cycle*—the cycling of water among the atmosphere, land, and oceans through precipitation and evaporation—will intensify in a warming planet. A warmer climate will enhance rates of evaporation and precipitation for the globe as a whole (Wetherald and Manabe 2002). Regional precipitation distributions may be drastically altered, leading to an increase in the intensity and frequency of rainfall in some regions and to pervasive drought in other regions. In addition to changes at the global level, local dynamics may influence intensity and frequency of precipitation as well (Seager et al. 2012).

The relationships between changes in the Arctic and climate change impacts are continuously being addressed; the dynamics are complex. Overall, as with many other phenomena in climate change, trends are consistent with major shifts despite periodic changes, for example, in the amount of ice that survives the summer melt (Overland et al. 2012; Samenow 2012; Samenow 2013; Gillis 2013a).

Changes in the Arctic due to climate change could cause weather systems to stall more frequently as they move across North America and Europe. Differences in temperatures between the Arctic and mid-latitudes affect the location and speed of the northern polar jet stream, which drives weather systems from west to east across the Northern Hemisphere. Because the Arctic is warming faster than the mid-latitudes, the difference in temperature is decreasing and the jet stream is slowing down. Also, the jet stream is developing large southward bends more frequently. The changes may increase the chances and duration of extreme weather in the Northern Hemisphere (Francis and Vavrus 2012).

Climate models predict a poleward shift in the storm track and descending branch of the Hadley circulation. The descending branch is currently near 30 degrees north and south of the equator, and is characterized by a downward movement in air that creates high pressure and limits cloud and precipitation for much of the year. Going forward, the climate models predict this downward flow of air will occur further northward in the Northern Hemisphere, thereby reducing precipitation across much of the southern tier of the United States and southern Europe. Increased evaporative losses due to rising temperatures are projected to increase aridity in the southwestern United States (Seager et al. 2007). This is expected to reduce runoff and reduce water availability for the southwestern United States, which may pose challenges for growing urban areas such as Las Vegas, Phoenix, and Los Angeles. Climate model simulations already suggest that some of the changes in the water cycle observed since 1950 can be attributed to anthropogenic influences. However, an increase in global mean precipitation is projected due to an intensified global water cycle in a warming planet, corroborating paleoclimatic studies that show that the Earth has been wetter during past warm epochs and cooler during past dry epochs. Despite an overall warmer and wetter planet, changes in precipitation are projected to be nonuniform. Significant increases in precipitation are projected at high latitudes and near the equator with decreases along the poleward edge of the subtropics near 30–35 degrees latitude (IPCC 2013, SPM-16). Thus, many regions in the arid and semi-arid mid-latitudes will likely experience less precipitation, while regions with typically abundant moisture will likely receive more precipitation (IPCC 2013, SPM-16). While the IPCC has high confidence in overall patterns of precipitation change, uncertainty remains regarding the magnitude of the changes.

Warming will result in additional challenges for water availability in mountainous areas where water is stored in the snowpack, a defining feature of states in the western United States. Observations have shown that a 1°C (1.8°F) warming has significantly altered the fraction of precipitation stored in mountain snowpack in the transitional elevations of the Sierra Nevada and Cascades since 1950 (Mote et al. 2005; Abatzoglou 2011). Apart from changes in precipitation, warming not only decreases mountain snowpack storage, but also changes the timing and magnitude of downstream water availability (Elsner et al. 2010; Cayan et al. 2008) and hydroelectric capacity (Hamlet et al. 2010). These changes—coupled with increased water demand associated with warmer temperatures, preexisting water allocations and rights, and inherent climate variability—are likely to make water an even more coveted resource, which may result in water conflicts in already stressed systems (Karl et al. 2009).

Some areas may see a decrease in water availability, with subsequent consequences for populations and the economy; however, other areas may be in for too much of a good thing. An intensification of the hydrologic cycle, coupled with an increase in the water-holding capacity of the atmosphere, has and will continue to increase the potential for extreme precipitation events (Min et al. 2011). Such changes not only increase potential flooding hazards but may also increase potential contamination of drinking water. For example, flooding in cities may overwhelm municipal water utilities, resulting in sewer overflow and standing water that is susceptible to high concentrations of *Giardia, Cryptosporidia,* and coliforms and vector-borne infections (as well as mosquito populations).

Tropical storms have been at the forefront of media attention after the record-setting summer of 2005 in the tropical Atlantic that featured fifteen hurricanes, four of which reached Category 5 status (winds exceeding 155 miles per hour). Although there appears to be no global trend in tropical storm frequency, the number of major hurricanes (Categories 4 and 5) has nearly doubled in the last thirty-five years (Webster et al. 2005). This observation is consistent with the increase in tropical surface ocean temperatures over the last fifty years. Although we cannot attribute a single hurricane or season to anthropogenic climate change, a warming of the tropical ocean would provide greater energy to fuel more powerful tropical storms and hurricanes.

In its Fifth Assessment the IPCC, after internal discussion regarding use of outlier scientific findings, predicted that sea levels will continue to rise during the twenty-first century. Under the different scenarios that the IPCC uses based on assumptions of various kinds, the rate of rise "will *very likely* exceed that observed during 1971–2010 due to increased ocean warming and increased loss of mass from glaciers and ice sheets" (emphasis in original; IPCC 2013, SPM-6). Over the period 1901–2010, global mean sea level rose by 0.19 (0.17 to 0.21) meters. (IPCC 2013, SPM-6). In its previous report, in 2007, the IPCC predicted that sea levels would increase 0.2 to 0.6 meters (0.92 to 1.40 feet) by 2100, compounding the effects of tropical storms on low-lying infrastructure (Alley et al.

2007). These estimates are considered very conservative. Scientists are looking to evidence from ancient coastlines to improve estimates of sea level rise due to climate change (Gillis 2013b).

As modeling of sea level rise matures, estimates of possible sea level rise by 2100 have been revised upward (Rahmstorf 2013a, 2013b). Informed by advances in sea level estimates, a 2012 NOAA study finds a greater than 9 in 10 chance that "global mean sea level will rise at least 0.2 meters (8 inches) and no more than 2.0 meters (6.6 feet) by 2100." To help decision makers reduce the vulnerability of coastal infrastructure while taking uncertainty into account, the study provides four scenarios. The study recommends the highest scenario, showing 2 meters of global sea level rise by 2100, for decisions on the construction of new power plants and other long-lived infrastructure. In addition, the study cautions that local changes in sea level will vary from place to place. In part, this is due to the fact that wind patterns and ocean temperatures vary around the planet, causing sea levels to rise in some areas more than in others (Emanuel 2012). In addition, sea level rise may vary due to local and regional conditions. For example, coastal areas are moving upward in Alaska and the Pacific Northwest, but subsiding in the Mississippi Delta (Parris et al. 2012, 1–3).

Sea level rise of 1 meter or more would displace hundreds of millions of people in low-lying coastal areas in Asia and inundate coastal cities, including London, Bangkok, Miami, New York, and Cairo. Low-lying areas in Florida and the US Gulf Coast would be particularly vulnerable to pressures from sea level rise, increased storm strength, and receding river deltas. The approximate 1-foot rise in sea level in New York since 1900 contributed to the record-setting storm surge experienced with Hurricane Sandy in October 2012. If current trends continue, members of human settlements affected by rising sea levels would be displaced and become "climate refugees."

The 2013 draft National Climate Assessment Development Advisory Committee (NCADAC) report warns that rising sea level, which is projected to rise by another 1 to 4 feet in the current century according to some models, and by as much as 6.6 feet by 2100 according to risk-based analyses, has the potential to negatively impact the nearly five million Americans who live within four feet of local high-tide levels. Coastal infrastructure—including roads, bridges, rail lines, energy infrastructure, and port facilities such as naval bases—is and will be increasingly at risk from storm surges, which are exacerbated by rising sea levels. Salt marshes, mangrove forests, barrier islands, and reefs defend coastal ecosystems and infrastructure, such as transit systems and buildings, against storm surges. The NCADAC report points out that climate change and human-caused ecosystem and landscape modifications often increase the vulnerability of these systems to damage from extreme events, while also reducing their natural capacity to mitigate the impacts of those events.

In addition, climate change models suggest there may be changes in the location and timing of *upwelling*. Upwelling provides nutrients for phytoplankton, which provide food for other marine life (Warner and Schofield 2012). Changes in the location and timing of upwelling of nutrients from ocean depths would cause changes in fisheries, which are already under pressure from overfishing in many locations. For low-lying island and coastal areas, changes in upwelling combined with sea level rise, ocean acidification, and vulnerability of coral reefs may create pressure for changes in maritime jurisdictional zones and other policies to better protect fisheries and livelihoods (Warner and Schofield 2012).

The direct effect of climate change on human health has been widely debated, as warming alone is projected to bring both detrimental and beneficial effects to humans. As heat-related mortality increases significantly when the air temperature exceeds 32°C (90°F) (Davis et al. 2003), warming is projected to result in heightened health risks for much of the world's population (Adger et al. 2007). Recent excessive heat waves, such as those that ravaged much of Europe in the summer of 2003 and western Russia in 2010, resulted in over 60,000 heat-related deaths. It is important to note that these heat waves may not have been directly caused by climate change, but rather by a persistent weather regime, although some recent work suggests a likely climate change link. Studies suggest that human-caused climate change loads the dice in favor of such events, making them far more likely (e.g., Stott, Stone, and Allen 2004; Rupp et al. 2012). Increases in temperature will allow heat waves to become more intense and longer-lasting (Diffenbaugh and Ashfaq 2010). Models also suggest that extremely warm summers, like the summer of 2003 across much of Europe, the summer of 2010 across western Russia, and the summer experienced across much of the United States in 2012, will be commonplace by the mid-twenty-first century. On the other hand, warming is projected to decrease cold-related mortality in northern Europe (Christidis et al. 2010). However, the world's least-developed countries lack the resources to cope with extreme temperature and are more susceptible to extreme heat. The IPCC has thus concluded that the net impacts of warming would increase human mortality.

Under such conditions, the 2013 draft NCADAC report cautions, new health threats will emerge and existing health threats will intensify. In addition, the report states that climate-change impacts "are expected to be most difficult for those with fewer resources to adapt. Some changes will be disruptive to society because our institutions and infrastructure have been designed for the relatively stable climate of the past, not the changing one of the present and future" (NCADAC 2013, draft).

Climate projections suggest that high latitudes will experience the greatest rate of warming over the next century. Warming at high latitudes may be beneficial for some communities, improving prospects for agriculture and timber harvests and ice-free shipping ports. However, the effects of warming will not be as kind for communities whose culture and way of life depend on ice. Changes in climate over the last few decades have already hurt Inuit villages. Many Alaskan coastal areas have experienced dwindling food supplies and infrastructure damages as reduced sea ice and melting permafrost have made villages vulnerable to erosion. As a result, coastal communities have relocated inland, while animals dependent on sea ice are in decline and are projected to experience continued declines on a warming planet (e.g., Durner et al. 2009).

The village of Newtok, Alaska, is surrounded on three sides by the Ninglick River as the river journeys to the Bering Sea. The Ninglick has steadily been consuming Newtok's land. From 1954 to 2003, erosion occurred at an average rate of 68 feet per year in the area of the village. (Climate Adaptation Knowledge Exchange 2010). The river and the erosion that is consuming Newtok's land have both been moving at extraordinary speed during recent years as a result of rapid, climate change-induced ice melt. A report by the US Army Corps of Engineers predicts that the highest point in the village might be underwater as early as 2017 and states it will not be possible to protect the village in its current location. Eventually, the village must be abandoned; the residents, whose forebears have been fishing and hunting in the region for centuries, will have to leave their ancestral lands, becoming America's first climate change refugees. If the village cannot be relocated, the 350-person community will be dispersed among the villages and towns of Alaska and to points unknown. The residents of Newtok are not alone: more than 180 native communities in Alaska are currently being flooded and losing their ancestral lands as the warming climate melts the Arctic ice (Goldenberg 2013).

Global warming's effects extend across the sectors of human health, the economy, politics, and international relations. Climate change has been projected to worsen urban air pollution (e.g., Hayhoe et al. 2004); increase malaria and other infectious diseases (Lafferty 2009); affect energy supply and demand (Miller et al. 2008; Sathaye et al. 2013) and transportation performance; and challenge regional and local economies. While some individuals and countries have resources to avoid many of the adverse effects of climate change, others may not be so resilient.

Wealthier nations and people may be able to lessen or adapt to the effects of climate change; however, the negative consequences of climate change fall disproportionately on poor and vulnerable nations and populations. According to the IPCC, the regional effects of climate change will vary over time and will be determined in part by the ability of different systems—both societal and environmental—to mitigate or adapt to climate change (NASA 2013).

Effects on Natural Systems

Ecosystems face numerous challenges brought about by climate change. Scientists can better understand and predict how future climate change is likely to impact natural systems by examining ecological responses to historical climate change using paleorecords such as tree rings and lake sediments. These repercussions take the form of both slow, long-term transitions in the biosphere and sudden, rapid ecological disturbances that radically shift the ecological make-up of local biota.

Marine ecosystems are starting to show the considerable negative effects of rising global ocean temperatures and ocean acidification. The ocean has been a sink of carbon dioxide, taking up approximately a third of the additional carbon emitted into the atmosphere since 1850. Over the last 250 years, oceans have absorbed 530 billion tons of CO_2, triggering a 30 percent increase in ocean acidity (National Resources Defense Council 2009).

This additional carbon translates into more dissolved carbon dioxide that alters ocean chemistry and makes it difficult for species to build shells and skeletons from calcium, especially in early life stages. Such changes affect oysters, some types of plankton, and other species important to marine food webs in coastal areas. Ocean acidification also slows photosynthesis in some species. It is still unclear to what extent key species in coastal and high-latitude marine ecosystems will be able to adapt to more acidic conditions, especially areas facing multiple difficulties such as warming waters, arrival of new predators, and pollution (National Research Council 2010). In addition, nearly half of the globe's coral reefs, home to the most biodiverse marine ecosystems on earth, are threatened by the combination of ocean acidification and rising temperatures. Research off the coast of Washington shows some organisms are adapting better than others, shifting the balance of predators and prey in the local ecosystem as ocean acidity increases (Wootton, Pfister, and Forester 2008).

The effects of climate change on terrestrial ecosystems are also of grave concern, especially those that provide considerable ecosystem services to humans and have additional aesthetic and spiritual values. Species generally fall into two categories: generalists, which are able to span broad ranges of climate conditions and are thus more resilient to changes in climate, and specialists, which generally exploit a very specific ecosystem niche and are often constrained geographically to specific areas. In light of these constraints, many of these specialists are adversely impacted by even small changes to the ecosystem, and are subsequently classified as threatened or endangered.

Paleorecords spanning thousands of years reveal that changes in vegetation composition across the landscape mirror changes in temperature, precipitation, and atmospheric composition. Long, slow changes in climate generally impact species by geographically altering what are called *bioclimatic envelopes:* the range of average temperature and precipitation conditions under which that species can continue to reproduce and sustain a viable community. It is widely held that warming will generally "push" the range of many species up in both latitude and elevation (e.g., Loarie et al. 2009), although changes in precipitation and moisture availability may drive more complex changes in species whose geographic range is more dependent upon moisture than on temperature (Crimmins et al. 2011, Williams et al. 2010, Lutz et al. 2010).

These shifts in species range are difficult for humans to observe as they take place over multiple decades, but changes in site-specific, local communities are more obvious. In Alaska, where significant warming has been observed over the past half-century, there has also been an expansion of the boreal forest to higher elevations, an increase in the density of shrubs on the tundra, and a drying pattern that has converted freshwater ponds and wetlands into dry, grassy meadows (Chapin et al. 2005, Riordan et al. 2006). Vegetation communities in any location naturally grow and change over time, in a process called *succession*. But climate change puts succession on a different trajectory, one that has a different end state, or "climax" stage. These two different vegetation community trajectories have different sets of faunal species associated with them, since the food chain starts at the lowest level with the herbivores that depend on specific plants for their diet. In the Alaska example, warming facilitates the uphill migration of the maximum tree line elevation and the expansion of the boreal forest habitat. Increases in shrub density on the tundra may provide more habitat for small mammals, but be less supportive for the large mammals with hooves (like caribou) that prefer low-to-moderate–density vegetation habitat. Meanwhile, the decrease in wetlands across the Alaskan interior reduces habitat available to the moose, beaver, and other water-loving species that forage or live in the wetlands.

Climate change alters the process of succession through two paths within the successional cycle of an ecosystem: competition and disturbance. There is competition for every available nutrient and resource on the landscape, and plants and animals evolve to take advantage of certain conditions in order to out-compete other species. Some reproduce at the highest rate, while others have a life cycle that capitalizes upon certain climatic conditions. For example, the spring bloom in the Northern Hemisphere has recently arrived between four and eight days earlier than it did during the 1950s due to warming (Schwartz, Ahas, and Aasa 2006). Differences in how species respond to climate change may potentially result in a mismatch in timing between consuming species and food resources. For example, changes in the extent of snow cover weaken the ability of snowshoe hares to camouflage themselves and avoid predation (Mills et al. 2013). Floral species that depend on animal and bird seed consumption and elimination in order to reproduce suffer a reduction in their reproductive potential due to the absence of the migrating fauna, while other species that require winds to disperse their seed may capitalize on an opportunity to expand. This type of pattern is mirrored in the case of changes in precipitation; some species capitalize on drought conditions, while others take advantage of wetter conditions. This sort of opportunism is one of the factors that encourage the spread of invasive species, which capitalize on "new" climate conditions as the existing species struggle to adapt. Buffelgrass (*Pennisetum ciliare*), a species native to much of subtropical Africa and Asia and introduced as a livestock forage in the southwestern United States and Mexico in the 1930s, has outcompeted native vegetation and expanded its range in recent years as a result of warming conditions and abnormal years of precipitation. Buffelgrass is highly flammable; as a result of its spread, wildfires have become increasingly prevalent within the fire-intolerant ecosystem across the Sonoran desert. Projected changes in climate are likely to further benefit the invasion of such annual grasses and threaten desert and rangeland ecosystems across the southwestern United States (Archer and Predick 2008).

The other path by which climate change affects ecological communities is through periodic disturbance. Ecological disturbances such as fire, flood, avalanches, insect infestations, severe weather events, disease, and regional die-off occur naturally in almost all ecosystems; these disturbance events "reset the clock" on succession processes and have occurred for millennia. Some species are not only adapted to disturbance, but require it to sustain the population, such as giant sequoia trees that only sprout when a low-intensity fire opens their cones. The frequency and intensity of these disturbance events, however, is dramatically altered by climate change, to the detriment of species that are less resilient or adaptable. Wildfires are a regular occurrence across much of the western United States, and most forest and grassland ecosystems there are adapted to regular wildfire events. However, the projected warmer and drier summers across regions like the northern Rocky Mountains by the middle portion of the twenty-first century will significantly increase the possibility of larger and more frequent fires (Brown, Hall, and Westerling 2004; Westerling et al. 2011b). This type of climate-induced change in the wildfire regime favors the expansion of species that are adapted to more frequent fires (like ponderosa pine trees) over the existing species that cannot resist fire and take longer to reestablish (such as lodgepole pines and spruce trees). In the southwestern US deserts, projected increased fire danger has the potential to combine with an ongoing expansion of annual invasive grasses in a self-perpetuating worsening cycle that will eventually convert deserts currently populated by cacti and dry habitat shrubs into deserts covered by a homogenous invasive grassland (Abatzoglou and Kolden 2011), effectively eliminating the habitat of the threatened desert tortoise (*Gopherus agassizii*) and other species. In addition to its impact on ecosystems, wildfire affects humans via its direct impact on life and property, regional impact on air quality and human health, and the significant economic costs of fire suppression.

Threatened and endangered species like the desert tortoise face the greatest potential consequences from climate change. The most vulnerable species reside in novel climates: geographic locations that harbor a special climate, ultimately providing a unique habitat exploited by a specialist species. Often, these specialists are *endemic*, meaning they are only found in a specific location. Climate change will extinguish these novel climates and push vulnerable and endemic species toward extinction in two ways: 1) by pushing the climatic conditions at that location beyond a survivable threshold for the vulnerable species, thereby destroying the novel climate (Williams et al. 2007); or 2) altering the characteristics of ecological disturbance so that the species is unable to survive and adapt to the new disturbance. In the Arctic, the polar bear population has been reduced through climate change–related habitat alterations to the point where the species is now listed as threatened. Climate change has significantly reduced the polar sea ice, and warmer autumn months in the last decade have delayed the formation of sea ice along the coast by several weeks, at a time when polar bears are at greatest risk. In a cycle that worsens over time in response to previous events, the warming also triggers stronger autumn storms that both degrade shoreline habitat for polar bears and their food sources, and make it even more difficult for sea ice to form. A 2007 US Geological Survey report (released prior to the record 2007 low sea-ice event) conservatively projected that polar bears will decline throughout all of their range (sea-ice habitat) during the twenty-first century and will be extinct within portions of their range within seventy years at the projected rate of warming (Amstrup, Marcot, and Douglas 2007).

Case Study: California

The state of California provides a suitable test bed for examining regional and local climate change repercussions because of the complex physical controls on regional climate across the state and the sensitivity of the regional economy to climate changes (Abatzoglou et al. 2009; Hayhoe et al. 2004). California is also the most populous US state and contributes about 6 percent of total US greenhouse gas emissions. Finally, it has exhibited a progressive stance on environmental policies pertaining to climate change, and is regarded as a leader to guide climate actions for other US states.

California has witnessed an increase in temperature of 0.8–1.5°C (1.4–2.4°F) over the last century, with warming trends generally more pronounced for minimum temperatures than maximum temperatures (Cordero et al. 2011). Results from a large number of modeling runs project that statewide temperatures will increase 1.5–4.5°C (2.5–8°F) over the course of the twenty-first century (Cayan et al. 2008). As of March 2013, the most recent set of climate models developed for the Fifth Assessment report of the IPCC concur with such projections. Models project the warming to be most pronounced during midsummer (up to 6.5°C or 11.7°F warmer), "enough to make many coastal cities feel like inland cities do today, and enough to make inland cities feel like Death Valley" (Hayhoe et al. 2004). However, other recent work suggests that climate change may increase coastal upwelling and lower temperatures near the California coast.

An increase in average temperatures means that heat waves will be more frequent and more intense. During July 2006, an extremely warm air mass stagnated over the western United States. Fresno experienced six consecutive days of 43°C (110°F) temperatures; the low temperature in Death Valley on July 24 was 38°C (100°F). An extended period of record-setting heat during the latter half of the month contributed to the death of over 160 people statewide (Gershunov et al. 2009). Models predict that heat waves of the same magnitude will be commonplace in California by the end of the twenty-first century

(Miller et al. 2008). Moreover, future heat waves in California are projected to have significant impacts in coastal areas, where most of the state's population lives. The coastal areas exhibit more humid conditions that inhibit overnight cooling. High temperatures and humid conditions have significant negative consequences for human health (Gershunov and Guirguis 2012). Increases in temperature, in conjunction with warming's effect on air quality, could indirectly increase respiratory and cardiovascular problems that currently contribute to approximately 8,800 deaths and over a billion dollars in health care costs annually in California (Karl et al. 2009). However, there are steps we can take to avoid this outcome. If more stringent control measures are implemented, air quality may continue to improve in California, even with a warming atmosphere.

Predicted statewide increases in temperature and population will drive the demand for air conditioning and energy use, unless increased energy efficiency and self-generation can make up the difference. In California, the peak hour greatly exceeds the energy needs for the rest of the year. With expected increases in population and temperature, peak electricity demands may march upward and require changes in the electrical system to cope with additional stressors (Miller et al. 2008). However, the number of rooftop photovoltaic systems (which generate electricity from sunlight) is growing quickly, changing the shape of the net load curve for much of the state and increasing the need for fast-ramping energy resources (California Independent System Operator 2013).

Despite mixed projections of changes in annual precipitation totals for the state, warming will induce significant changes in hydrology as more precipitation in the Sierra Nevada falls as rain rather than snow and warmer temperatures accelerate snowmelt. Projections of the amount of water stored in snowpack on April 1st, which are often used to develop water supply projections for the summer, are expected to decline 32–79 percent by the late twenty-first century (Cayan et al. 2008). The trickle-down effect of changing water resources is likely to hurt the state's economy and vitality and damage a number of treasured ecological facets that define the state. First, warmer and wetter storms during winter are projected to increase winter runoff. Since present-day reservoirs cannot store anticipated increases in winter runoff for use during the dry season, peak floods will likely be more intense (Anderson et al. 2008, Das et al. 2011). Not only will flood impacts be worsened by rising sea levels, the combination of the two may result in salt-water intrusion to the Sacramento–San Joaquin delta, further threatening the state's water supply. Aside from such threats, operating costs to sustain water availability for the state under a warmer and drier scenario could increase nearly $500 million per year by the mid-twenty-first century (Medellin-Azuara et al. 2008).

A decline in water availability would damage California's multibillion-dollar agriculture industry. A combination of more winter floods, reduced water availability, and increased evapotranspiration (a combination of evaporation from water and soils and release of moisture from plants) during the growing season means that water-intensive crops are likely to suffer (Field et al. 1999; Wilkinson et al. 2002; Miller, Bashford, and Strem 2003; Cayan et al. 2011). Warmer temperatures and decreased soil moisture would dramatically increase irrigation demands and reduce crop yields. Warmer temperatures during winter may reduce the number of chill hours needed for fruit and nut trees to become dormant and set fruit, thereby limiting productivity in many areas (Baldocchi and Wong 2008). Likewise, the state's treasured wine industry is projected to be in jeopardy as warming across many world-class wine grape–growing areas would make them inhospitable to fine wine production (White et al. 2006, Diffenbaugh et al. 2011). However, the cooler climates of coastal California and the foothills of the Sierra Nevada may still be viable areas for wine grape production in a changing climate.

Climate change's ecological effects in California will be highly variable, since the state has arguably the greatest biological diversity in the contiguous United States, with numerous endemic and vulnerable species occupying hundreds of microclimate niches. Generally, however, climate change is expected to alter the timing of important ecological events, such as spring green-up and the onset of the bloom, while also shifting ecological zones northward and upward in elevation. Assuming that plants maintain ideal growing conditions under a temperature optimum, a 3°C (5.4°F) warming over the next century would require current vegetation distributions to migrate northward about 5 kilometers (3 miles) per decade (Loarie et al. 2009). The warming of waters in the Sacramento Delta is expected to increase the frequency of water temperatures lethal to delta smelt, an endemic species critical to ecosystem health (Cloern et al. 2011).

In the northwestern portion of the state, the consequences of warming and summer drying will reduce the viability of the temperate forest and make it susceptible to more frequent and more devastating wildfires (Westerling et al. 2011a). This is likely to help change the rainforest to a mixed-conifer forest not unlike that which carpets the eastern (lee) side of the coastal mountain range and the mid-elevation slopes of the Sierra Nevada. To the south, more frequent and severe wildfires have already been chipping away at the remnant stands of big-cone pine crowning the high peaks in the Big Sur region;

these stands will disappear entirely and be replaced by shrubs. The range of coastal redwoods may shrink due to warming and increased moisture stress (Johnstone and Dawson 2010). Invasive species, which have already expanded across much of coastal California, will capitalize on opportunities to fill these empty niches, to the detriment of native species.

Along the 400-mile length of the Sierra Nevada, the species populating the high alpine meadows and forests will be faced with a reduced snowpack and longer snow-free season. This may be an opportunity for some species of animals and birds. Other species, however, may experience die-off or increased competition as temperatures rise and moisture declines. A lengthening of summer moisture-limiting conditions will be detrimental to the array of wildflowers that are accustomed to ever-moist soils in the alpine meadows, and the bird and insect species that depend heavily upon this flora will see a reduced food source. Further down the slope, species may move depending on their primary requirement (temperature, moisture); the diverging movement will disrupt current plant community dynamics and favor those that can adapt quickly.

In the deserts of eastern California, the situation is already dire. Disturbance in deserts most often takes the form of rare and intense rainfall events that unleash floods across a landscape with little vegetation to hold it together. Invasive annual grasses have covered much of the Mojave Desert and northeastern California's Great Basin over the last two decades, replacing native xeric (dry habitat) shrubs such as black brush, creosote bush, and sagebrush. One of the primary drivers of this replacement has been increasing wildfire, which was relatively rare prior to European settlement. Now, however, wildfire occurs in a positive feedback loop with invasive grasses: occurrence of one increases the occurrence of the other. Desert species are highly specialized to deal with long periods of drought followed by short, infrequent rainfall events and do not quickly recover from more frequent disturbances. Nor do they adapt well to these types of rapid changes; there is considerable evidence that such iconic species as the Joshua tree, with its poor dispersion capacity, will disappear because of climate change (Cole et al. 2011)

Climate change–induced alterations will be particularly acute for refuges and corridors for particular species. The wildlife refuges that dot the Sacramento Valley as stopovers for several hundred bird species that migrate over the state each year may dry up, and no similar or equal location exists further north, effectively eliminating this refuge. Existing habitat corridors for migrating elk and mule deer in the Eastern Sierra are likely to be left empty of these large mammals, while they struggle to find new corridors further upslope and northward amid private, unregulated lands. In a state as populous and developed as California, species may be unable to move onward and upward, or into more suitable locations, to adapt to climate change. Millions of Californians have built the state into a network of development that creates a formidable roadblock to finding new, more suitable habitat.

The California coastal waters are rich in marine life attracted by the upwelling of nutrient-rich deep water (Bograd et al. 2009; NOAA 2012). The upwelling waters are also rich in CO_2. As a result, researchers are concerned that California waters may soon experience high levels of acidification. A study completed in 2012 suggests that ocean acidification will lead to a substantial drop in seawater calcium needed for shellfish, small marine snails, and other marine life in California coastal waters within the next twenty to thirty years. It is not clear how California marine life will respond to this change along with concurrent changes expected from warming ocean temperatures and reduced dissolved oxygen levels (Gruber et al. 2012). The oyster industry off the coast of California and neighboring states has already seen a steep decline linked to ocean acidification. Awareness of this link has led to opportunities for adaptation in the oyster industry (Weiss 2012).

Case Study: New York

The state of New York provides an additional case study and interesting contrast in climate change and its projected consequences on the opposite side of the country. Climate-change effects in New York in many instances will present different concerns, or different levels of concern, from those experienced in California. For example, while California is more threatened by drought and ocean acidification along its extensive shoreline, New York is more vulnerable to sea level rise, powerful rainstorms, and storm surges. While hurricanes and associated storm surges are of great concern to the people of New York City, hurricanes do not impact California and storm surges are of less concern along California's coastlines.

A report written by fifty scientists from Cornell University, Columbia University, and the City University of New York was commissioned and released by the New York State Energy Research and Development Authority. The report, which examined recent weather patterns in comparison with historical patterns, predicts that average annual temperatures in New York State will rise by 4°F to 9°F by 2080. Heat waves will become more frequent and intense, and the state's air quality will worsen (Rosenzweig et al. 2011).

A National Resources Defense Council (NRDC) report states that twenty-three counties in New York now experience unhealthy smog levels that will worsen in a warmer climate. Increased heat, particularly in combination with poor air quality, will result in increases in human heat-related illnesses and deaths. Heat-related mortality in the state is projected to rise by 70 percent by mid-century. In addition, ragweed pollution is currently a problem in thirty-seven counties in the state; climate change will result in an increase in pollen production by plants, an increase in the seasonal duration of various types of allergen production, and an associated increase in allergies and asthma (NRDC 2013).

Other sectors, such as energy production, will also be challenged or overwhelmed by the increase in heat. Electrical demand is expected to increase significantly during the warmer months as people try to combat the heat by cooling residences, businesses, and public buildings. (Rosenzweig et al. 2011).

The NRDC report also predicts that waterborne illnesses and infectious diseases like dengue fever, West Nile virus, and Lyme disease will increase. A warming climate will also result in toxic algal blooms in bodies of freshwater and along coastal shorelines (NRDC 2013); this will further threaten human health and will lead to the death of aquatic species as well.

Increasing winter temperatures may shorten the duration of Great Lakes ice cover, permitting more moisture to rise from the lakes. The additional moisture would then fall as snow in western New York during cold weather. While the severity of winter snowfalls is likely to increase, the number of days with snow on the ground is likely to decrease. In coastal regions and estuaries, nor'easter blizzards may more often turn into severe rainstorms. Precipitation is predicted to rise by as much as 5 to 15 percent and will often occur as heavy downpours, leading to flooding and its associated impact on water resource quality and agriculture, as well as infrastructure such as transportation systems, buildings, water treatment plants, sewer systems, and roadways and bridges. Much of the increase in precipitation is anticipated to occur during the winter months (Rosenzweig et al. 2011). The increase in flooding and anticipated increase in sewer system failures due to extreme weather events is expected to threaten human health in seventy counties within the state (NRDC 2013).

While the winter months will receive more precipitation, less rainfall is predicted for the summer months. As a result, summer droughts are expected to increase in duration and intensity. Drought-related impacts will further affect water supplies, agricultural production, natural ecosystems, ecosystem processes, and energy production in the state (Rosenzweig et al. 2011). Summer droughts are expected to result in water shortages in 26 percent of New York's counties by mid-century (NRDC 2013).

As the climate changes, plant, animal, reptile, and bird populations are expected to crash. Species will shift their ranges in response to changes in their habitats and food supplies, or will attempt to do so. Many species will be unable to make the transitions quickly enough, or will be unable to find enough suitable habitat in other locations. Invasive insects and weedy species are expected to increase. This will result in wide-ranging effects on natural systems as the changes cascade through interrelated species' food webs.

As stated in the state's 2010 Interim Climate Action Plan, New York is already experiencing warmer temperatures, especially in winter, leading to shorter periods of winter snow cover and a decline in winter sports and tourism. Climate change is likely to result in significant agricultural-sector economic costs in New York from decreased crop yields and increased heat stress–induced decline in dairy cow milk production (New York State Department of Environmental Conservation 2010).

In addition, climate-change effects on the state's sugar maples may even threaten the survival of this iconic resource (Rosenzweig et al. 2011). In addition to being the primary source of maple syrup, sugar maples are responsible for much of the beautiful fall foliage that attracts visitors from around the globe; if this resource is lost, fall tourism is also likely to suffer.

Changes in water quality and abundance will also lead to decreased recreational and commercial fishing (New York State Department of Environmental Conservation 2010). This will represent an additional impact to the tourism and sport fishing industries. Meanwhile, increases in sea level will result in the loss of coastal wetlands and the species that depend upon them during all or part of the life cycles. As the oceans continue to warm, higher water temperatures will lead to declines in fish and shellfish populations. Already, lobsters and some other cooler-water marine species are shifting their ranges northward and out of the state, while the warmer waters are favoring the increase in numbers of species like the blue claw crab (New York State Department of Environmental Conservation 2010). Sea level rise will also promote saltwater intrusion in some areas.

Some economic costs associated with climate change—while just as real—are more difficult to quantify. How do we put a price on the heritage value of alpine forests or their function as habitat for endangered species? Moreover, some costs cannot be accurately calculated due to the unpredictable nature of future changes. On the one hand, climate change in other parts of the state, including in the Snow Belt, may bring about downturns in snow-based tourism. On the other, increased economic opportunity may arise as a result of a more moderate upstate climate and the additional availability of water resources. As Ackerman (2009, 54) states, "There is no hope of coming up with a single dollar amount that adequately summarizes the full range of climate impacts; too many of the impacts are incapable of being measured in monetary terms. Yet even without an impossibly comprehensive summary number, there are ample grounds for taking action to reduce climate damages."

In 2013, Governor Cuomo and the state of New York began to caution bond investors that climate change may pose a significant long-term risk to the state's financial well-being. Citing Hurricane Sandy and Tropical Storms Irene and Lee, the caution to investors is included on the bonds along with warnings about additional risks that could adversely impact the state's finances, including federal spending cuts, litigation against the state, and potential labor negotiations. According to Richard Azzopardi, a spokesman for the governor's office, "The state determined that observed effects of climate change, such as rising sea levels, and potential effects of climate change, such as the frequency and intensity of storms, presented economic and financial risks to the state. ... The extreme weather events of the last two years highlighted real and potential costs from extreme weather events, including the need to harden the state's infrastructure and improve disaster preparedness, both of which have been a priority of the governor" (Kaplan 2013).

The Interim Climate Action Plan (New York State Department of Environmental Conservation 2010) reports that global sea level rise, compounded by local subsidence of coastal land, have resulted in a local increase in sea level of 3 cm (1.2 inches) per decade in New York. By 2100, relative to the 2000–2004 baseline, sea levels are expected to rise anywhere between about 30 cm and 58 cm (12 and 23 inches) in the New York

City coastal areas and along the Hudson River estuary. When rapid glacial ice melt is factored into the models, sea levels are projected to rise almost 140 cm (55 inches) in the Lower Hudson Valley, New York City, and Long Island. Sea level rise will immensely exacerbate existing risks to coastal populations and lead to permanent inundation of low-lying areas, increased coastal and beach erosion, an increase in the frequency of storm surge–related flooding, and contamination of the Hudson River, coastal estuaries, and groundwater-based freshwater supplies with salt water. High water levels, heavy precipitation, and strong winds from fierce coastal storms already result in billions of dollars in damages, disrupt transportation and power production or distribution systems, and dramatically alter the barrier islands that help to protect New York City and the coastline from powerful storms and storm surges. Rising sea levels will worsen storm surge-related flooding, threaten vulnerable energy production and telecommunication facilities in coastal areas, and inundate coastal marshes, wetlands, and estuaries.

Judging from the impacts of Hurricane Sandy in 2012, which caused billions of dollars of damage (FEMA 2013), a sea level rise of even 23 inches could be devastating for the city of New York, damaging or destroying natural areas, buildings, roadways and bridges, rail and subway lines, tunnels into and off the island of Manhattan, utility and telecommunications networks, wastewater treatment facilities, recreational facilities, and other forms of infrastructure. A 55-inch increase in sea level could inundate large swaths of the city and leave noninundated areas of the city increasingly vulnerable to coastal storms and storm surges.

Sea level rise, higher temperatures, extreme weather events, and increases in precipitation, flooding, and coastal erosion will put transportation infrastructure and operations at risk. Asphalt pavements; other road, bridge, and runway surfaces; railroad tracks; and electrical wiring and conduits may be damaged by more frequent and extreme temperatures and by intense storms. Air conditioning requirements in transportation vehicles and in tunnels will increase, placing greater demand on energy systems. High winds during intense storms may result in more frequent temporary closures or restricted use of larger bridges (Rosenzweig et al. 2011).

Flooding, washouts and erosion, and mudslides or landslides will also adversely impact roadways and railroads running along inland rivers and streams, in areas where existing drainage is insufficient to cope with intense storms, or in areas of steep terrain. All transportation systems in the coastal regions are threatened by coastal storms and related storm-surge flooding hazards. In New York, some of these systems are located at low elevations along bodies of water. Some subways, railroads, and highways tunnel beneath the city below sea level. Sea level rise, extreme storm events, and storm surges will dramatically increase the probability of flooding. Sea level rise will eventually inundate low-lying transportation systems permanently unless costly mitigation or adaptation measures are taken. The increase in storm intensity will affect air transportation, increasing the number of delayed or canceled flights, temporarily shutting down airports, and detouring flights to other airports. Because hotter air provides less lift, airport runways may have to increase in length (Rosenzweig et al. 2011).

Intense storms increase and redistribute sediment loads in rivers, harbors, and shipping lanes, potentially increasing the need for dredging operations. Bridge foundations in some rivers are also scoured by *sediment transport* (the movement of various-sized particles due to the combined effects of gravity and the movement of the fluid in which the particles are located) during strong storms and heavy water flows. Yet sediment transport in New York waterways is not clearly understood, and sediment transport under future climate conditions is even less well understood. Sea level rise may dominate over time, reducing the need for dredging operations. On the other hand, sea level rise may also reduce bridge clearances over waterways below the limits set by the US Coast Guard or by other jurisdictions (Rosenzweig et al. 2011).

Increased evaporation in the Great Lakes or along the St. Lawrence River Seaway under severe and prolonged droughts and extended heat waves is likely to lower water levels enough to impede shipping. However, a warming climate should prolong the ice-free shipping season on the lakes and seaway, making them less prone to the ice floes or shore-to-shore freezes that have occurred in the past (Rosenzweig et al. 2011).

For the telecommunications sector, the greatest threats posed by climate change will result from an increase in short-duration but intense downpours, or large-scale weather events such as hurricanes that bring high winds, lightning, and flooding. These types of events are predicted to increase statewide. Nor'easters and winter precipitation such as freezing rain, ice, and heavy snow also impact telecommunication systems. As noted above, these events may lessen in many parts of the state and may worsen in areas subject to lake-effect weather systems (Rosenzweig et al. 2011).

Impact Avoidance: Adaptation, Mitigation, and Geoengineering

Future changes in climate have been portrayed as cataclysmic, which results in an abstractness that is difficult to deal with.

What can we do in the face of such predictions? We can accept the fate of dire outcomes, or devise ways to counter the detrimental effects of change. Three fundamental means have been proposed to temper detrimental impacts: adaptation, mitigation, and geoengineering.

Adaptation involves using climate projections to reduce risks associated with a changing climate. For example, low-lying areas vulnerable to sea level rise are developing strategies to minimize socioeconomic loss (Frazier et al. 2010). Another strategy involves managing freshwater resources, particularly in regions vulnerable to water shortages (e.g., Roy et al. 2010). In addition to precipitation changes, warming in snow-dominated areas is likely to alter the seasonal availability of water and force water managers to develop new approaches to optimize water resources (e.g., Milly et al. 2008). Likewise, adaptation strategies are being developed for agriculture (Lobell et al. 2008), species habitats and refuges (Rehfeldt et al. 2006), and forest disturbances (Littell et al. 2010).

Monitoring local variations in ocean acidity is making it possible for the oyster industry in the Pacific Northwest to adapt to ocean acidification. Another adaptation strategy is to protect and plant seagrass meadows and other marine plants to help offset ocean acidification locally by uptaking CO_2 from the seawater. A third strategy is to reduce other stressors, such as pollution, to help increase the resilience of coastal marine life at the local level (Ocean Carbon and Biogeochemistry Program, Ocean Acidification Subcommittee 2012).

The United Nations Environmental Programme (UNEP) has launched the Blue Carbon Initiative to protect coastal and marine mangrove, salt marsh, and seagrass ecosystems that provide carbon sequestration and other important ecological services (UNEP 2012). Protecting these systems helps local areas adapt to rising ocean acidity and helps mitigate global levels of anthropogenic greenhouse gases.

In its discussion on adapting to the impacts of climate change, the draft 2013 NCADAC report concludes:

Because of the influence of human activities, the past climate is no longer a sufficient indicator of future conditions. Planning and managing based on the climate of the last century means that tolerances of some infrastructure and species will be exceeded. For example, building codes and landscaping ordinances will likely need to be updated not only for energy efficiency, but also to conserve water supplies, protect against insects that spread disease, reduce susceptibility to heat stress, and improve protection against extreme events. ... Adaptation considerations include local, state, regional, national, and international jurisdictional issues.... Both "bottom up" community planning and "top down" national strategies may help regions deal with impacts such as increases in electrical brownouts, heat stress, floods, and wildfires. Such a mix of approaches will require cross-boundary coordination at multiple levels as operational agencies integrate adaptation planning into their programs (NCADAC 2013 executive summary, 6).

Mitigation involves efforts to reduce the amount of carbon dioxide in the atmosphere. Some anthropogenic change is already inevitable due to the long lifetimes of greenhouse gases and because ocean waters and feedback processes retain heat for long periods of time. Even if anthropogenic emissions stopped today, the planet would continue to warm for the next twenty to fifty years. Reducing further emissions will reduce the magnitude of climate change, the scope and scale of damaging impacts, and the costs of adaptation.

Efforts to reduce anthropogenic greenhouse gases are needed to keep the amount of warming to levels deemed tolerable to society and ecosystems. Mitigation can come through changes in the actions of individuals, companies, and nations. Cost-effective actions, such as consuming a diet based on chicken and vegetables as opposed to a beef-based diet, can make as much of an impact on personal carbon emissions as switching from driving a standard to a hybrid vehicle (Stec and Cordero 2008). Changes are also needed in the way we produce energy. In 2011, the IPCC stated that renewable energy could meet most of the world's energy supply by 2050 if it is supported by effective and efficient policies (IPCC 2011, 17–18, 21, 24–26). Demonstrating how to achieve this future, Germany is phasing out its nuclear energy facilities and shifting to renewable energy (Davidson 2012). Switzerland has stated it will phase out nuclear energy by 2034 (Mombelli 2012; swissinfo.ch 2013). Shifts away from nuclear are motivated by concerns over accidents, natural disasters, and the intractable problem of disposal of radioactive material. Nonetheless, some are calling for nuclear power as a major mitigator of emissions because the fission process itself does not produce greenhouse gases. Emissions of greenhouse gases, however, do result from the mining of fuels, building of facilities, and deconstructing of decommissioned nuclear power plants. Many remain adamantly opposed to nuclear energy. Energy efficiency, demand response, renewable energy (distributed and utility scale), storage, combined heat and power, and carbon capture and sequestration can decarbonize the energy sector without nuclear energy.

Greenhouse gas emissions can also be reduced through changes in agricultural processes. For example, sustainable biochar can be used as a soil amendment to improve soil quality and sequester carbon (International Biochar Initiative 2013a). A biochar protocol for the voluntary carbon-offset market is in review and is expected to be available for public comment after technical review is completed (International Biochar Initiative 2013b). Planting crops without tilling the soil, incorporating principles of conservation agriculture, is another method for reducing greenhouse gas emissions while improving crop resiliency (Food and Agriculture Organization of the United Nations 2010, 5–6).

Delayed enactment of mitigation forces us closer to temperature thresholds deemed detrimental to life on our planet. What happens when mitigation efforts fail, adaptation efforts are behind schedule, and severe consequences are imminent? Deliberately engineering the earth's climate system, or geoengineering, may be a last-ditch effort to counter warming. Thus far, geoengineering has primarily focused on methods to reduce the amount of solar radiation absorbed by the planet.

Osofsky and McAllister (2012) note that these methods generally fall into two categories: 1) solar radiation management, which involves increasing the planet's ability to reflect incoming sunlight; and 2) carbon dioxide removal, which involves the actual physical removal of carbon from the atmosphere in various ways. Solar radiation management measures work quickly to address the problem of warming, but do not address the underlying causes of the problem. Solar radiation management includes dispersing aerosols (or bright, reflective particles) into the stratosphere, putting into orbit a set of reflective solar shields, and deploying a fleet of unmanned spaceships to increase cloud formation in the subtropical oceans. Very danger-prone options, all. Increasing the reflectivity of the planet would theoretically reflect enough sunlight back to space to buffer anthropogenic warming. However, these methods would be costly and technologically difficult, would not necessarily prevent localized impacts, and could result in significant—and often unknown—safety and environmental consequences. For example, adding SO_2 particles to the atmosphere, as has been proposed, could further damage the ozone layer.

Carbon dioxide removal methods address the underlying causes of warming, but take effect much more slowly. The most commonly proposed measure includes the addition of micronutrients such as iron into the oceans to increase the capacity of phytoplankton to uptake carbon from seawater. As phytoplankton die, they sink to the seafloor, taking the excess CO_2 with them into the deep ocean and preventing the CO_2 from re-entering the atmosphere. Iron is plentiful on earth, and the cost of such a carbon removal scheme would be relatively moderate. However, iron-fertilization schemes have thus far yielded disappointing results in removal of atmospheric carbon, and numerous uncertainties abound regarding maximizing the effectiveness of ocean-fertilization schemes. Furthermore, these efforts could result in substantial and unpredictable alterations to marine ecosystems. Because phytoplankton form the basis for marine food webs, changes to phytoplankton populations could result in repercussions throughout these food webs and may also cause increased levels of methane and other heat-trapping gases as phytoplankton conduct natural metabolic processes and as they decay after death (Osofsky and McAllister 2012).

In short, geoengineering is another global experiment, similar to the ongoing experiment to increase greenhouse gas concentrations. While experiments that go awry in a lab can be quickly tended to, global experiments are not quickly reversible, and pose the risk of potential unintended outcomes. Not only would geoengineering be costly, the potential consequences could amplify climate alterations and introduce new ones (e.g., Robock et al. 2009). Geoengineering could cause additional conflict at the international level. It could aggravate rather than alleviate climate-related stressors. And, in the process, it could divert sorely needed resources away from adaptation and mitigation efforts. The community of nations needs to consider if and when geoengineering should be enacted, who will pay for it, and, ultimately, who controls the dial.

Conclusion

Climate change was recognized as a global environmental problem in the late 1970s, spurring an international scientific effort to understand its causes and effects. Climate models have been developed that allow researchers to provide climate projections well into the future. These projections can then be translated into consequences for both human and natural systems. Some changes have already begun, while others will be pronounced toward the middle and end of the twenty-first century.

Effects of climate change are continually under study. Some uncertainty remains as to how much the planet will warm and the associated changes in climate and extreme events. Some uncertainty comes from specific knowledge gaps, such as about our understanding of the mechanisms that control the rate of ice loss in Greenland and Antarctica (so scientists may more accurately predict the range of future sea level rise). Refinements in these projections will result from scientific advances that will better inform adaptation or mitigation strategies. We have a problem with no easy solution. Although carbon is a building block for life on earth, a carbon dioxide molecule emitted into the atmosphere remains in the atmosphere for twenty to one hundred years. Changes have been set in motion. The rate of change is a crucial component in how our planet weathers the storm. Efforts to slow the rate of warming via mitigation may provide additional time for effective adaptation. But mitigation challenges are beyond the scope of a single individual, or even a single country. Still, individuals and individual countries can lead by example.

References

Abatzoglou, J. T. 2011. Influence of the PNA on declining mountain snowpack in the western United States. *International Journal of Climatology* 31:1099–1256. doi:10.1002/joc.2137.

Abatzoglou, J. T., and C. A. Kolden. 2011. Climate change in western US deserts: Potential for increased wildfire and invasive annual grasses. *Rangeland Ecology and Management.* doi:10.2111/REM-D -09-00151.1.

Abatzoglou, J. T., K. T. Redmond, and L. M. Edwards. 2009. Classification of regional climate variability in the state of California. *Journal of Applied Meteorology and Climatology* 48(8):1527–1541.

Ackerman, Frank. 2009. *Can We Afford the Future? The Economics of a Warming World.* New York: Zed Books.

Adger, Neil, Pramod Aggarwal, Shardul Agrawala, Joseph Alcamo, Abdelkader Allali, Oleg Anisimov, Nigel Arnell, et al. 2007. Climate change impacts, adaptation, and vulnerability: Summary for Policymakers. In *Climate Change 2007: Impacts, Adaptation and Vulnerability. Contribution of Working Group II to the Fourth Assessment Report of the Intergovernmental Panel on Climate Change,* edited by M. L. Parry, O. F. Canziani, J. P. Palutikof, P. J. van der Linden, and C. E. Hanson, 7–22. Cambridge, UK: Cambridge University Press.

Alley, Richard, Terje Berntsen, Nathaniel L. Bindoff, Zhenlin Chen, Amnat Chidthaisong, Pierre Friedlingstein, Jonathan M. Gregory, et al. 2007. IPCC 2007: Summary for Policymakers. In *Climate Change 2007: The Physical Science Basis. Contribution of Working Group I to the Fourth Assessment Report of the Intergovernmental Panel on Climate Change,* edited by S. Solomon, D. Qin, M. Manning, Z. Chen, M. Marquis, K. B. Averyt, M. Tignor, and H. L. Miller. New York: Cambridge University Press.

Amstrup, S. C., B. G. Marcot, and D. C. Douglas. 2007. *Forecasting the Rangewide Status of Polar Bears at Selected Times in the 21st century: Administrative Report.* Anchorage, AK: US Geological Survey, Alaska Science Center.

Anderson, J., F. Chung, M. Anderson, L. Brekke, D. Easton, M. Ejeta, R. Peterson, et al. 2008. Progress on incorporating climate change into management of California's water resources. *Climatic Change* 87 (Suppl. 1):S91–S108. doi:10.1007/s10584-007-9353-1.

Archer, S. R., and K. I. Predick. 2008. Climate change and ecosystems of the southwestern United States. *Rangelands* 30 (3):23–28. doi:10.2111/1551-501X.

Baldocchi, D., and S. Wong. 2008. Accumulated winter chill is decreasing in the fruit growing regions of California. *Climatic Change* 87 (Suppl. 1):S153–S166. doi:10.1007/s10584-007-9367-8.

Bograd, Steven J., Isaac Schroeder, Nandita Sarkar, Xuemei Qiu, William J. Sydeman, and Franklin B. Schwing. 2009. Phenology of coastal upwelling in the California Current. *Geophysical Research Letters* 36:L01602. http://130.207.67.194/web_db_papers/pobex_pdfs/Bograd-2009.pdf.

Borenstein, Seth. 2013. US scientists report big jump in heat-trapping CO_2. Associated Press, March 5. http://bigstory.ap.org/article/us-scientists-report-big-jump-heat-trapping-co2.

Brown, T. J., B. L. Hall, and A. L. Westerling. 2004. The impact of twenty-first century climate change on wildland fire danger in the western United States: An applications perspective. *Climatic Change* 62:365–388.

California Independent System Operator. 2013. Fast facts: What the duck curve tells us about managing a green grid. October. http://www.caiso.com/Documents/FlexibleResourcesHelpRenewables_FastFacts.pdf, accessed November 23, 2013.

Cayan, D. R., E. P. Maurer, M. D. Dettinger, M. Tyree, and K. Hayhoe. 2008. Climate change scenarios for the California region. *Climatic Change* 87 (Suppl. 1):S21–S42.

Cayan, Daniel R., Susanne Moser, Guido Franco, Michael Hanemann, and Myoung-Ae Jones. 2011. Second California assessment: Integrated climate change impacts assessment of natural and managed systems. *Climatic Change* 109 (Suppl. 1):S1–S19.

Chapin, F. S., III, M. Sturm, M. C. Serreze, J. P. McFadden, J. R. Key, A. H. Lloyd, A. D. McGuire, et al. 2005. Role of land-surface changes in Arctic summer warming. *Science* 310:657–660.

Christidis, N., G. C. Donaldson, and P. A. Stott. 2010. Causes for the recent changes in cold- and heat-related mortality in England and Wales. *Climatic Change* 102:539–553.

Climate Adaptation Knowledge Exchange. 2010. Relocating the Village of Newtok, Alaska due to Coastal Erosion. By Kirsten Feifel and

Rachel M. Gregg. July 3. http://www.cakex.org/case-studies/1588. Accessed September 14, 2013.

Cloern, J. E., N. Knowles, L. R. Brown, D. Cayan, M. D. Dettinger, T. L. Morgan, D. H. Schoellhamer, et al. 2011. Projected evolution of California's San Francisco Bay-Delta-River system in a century of climate change. *PLoS ONE* 6 (9):e24465. doi:10.1371/journal .pone.0024465.

Cole, K., K. Ironside, J. Eischeid, G. Garfin, P. Duffy, and C. Toney. 2011. Past and ongoing shifts in Joshua tree support future modeled range contraction. *Ecological Applications* 21 (1):137–149. e-View. doi: 10.1890/09-1800.

Cordero, E. C., W. Kessomkiat, J. T. Abatzoglou, and S. A. Mauget. 2011. The identification of distinct patterns in California temperature trends. *Climatic Change* 108:357–382. doi:10.1007/s10584-011-0023-y.

Crimmins, S. M., S. Z. Dobrowski, J. A. Greenberg, J. Abatzoglou, and A. R. Mynsberge. 2011. Changes in climatic water balance drive downhill shifts in plant species' optimum elevations. *Science* 331:324–327.

Das, T., Dettinger, M., Cayan, D., and Hidalgo, H. 2011. Potential increase in floods in California's Sierra Nevada under future climate projections. *Climatic Change* 109 (Suppl. 1):S71–S94. doi:10.1007/ s10584-011-0298-z.

Davidson, Osha Gray. 2012. So far so good for Germany's nuclear phase-out, despite dire predictions. *Inside Climate News,* November 16. http://insideclimatenews.org/news/20121115/germany-ener giewende-nuclear-energy-fukushima-chernobyl-merkel-renewables.

Davis, R. E., P. C. Knappenberger, W. M. Novicoff, and P. J. Michaels. 2003. Decadal changes in summer mortality in U.S. cities. *International Journal of Biometeorology* 47:166–175.

Diffenbaugh, N. S., and M. Ashfaq. 2010. Intensification of hot extremes in the United States. *Geophysical Research Letters* 37:L15701. doi:10.1029/2010GL043888.

Diffenbaugh, N. S., M. A. White, G. V. Jones, and M. Ashfaq. 2011. Climate adaptation wedges: A case study of premium wine in the western United States. *Environmental Research Letters* 6:024024. doi:10.1088/1748-9326/6/2/024024.

Durner, George M., David C. Douglas, Ryan M. Nielson, Steven C. Amstrup, Trent L. McDonald, Ian Stirling, Mette Mauritzen, et al. 2009. Predicting 21st-century polar bear habitat distribution from global climate models. *Ecological Monographs* 79:25–58. doi:10.1890/07-2089.1.

Elsner, M. M., L. Cuo, N. Voisin, J. Deems, A. F. Hamlet, J. A. Vano, K. E. B. Mickelson, et al. 2010. Implications of 21st century climate change for the hydrology of Washington State. *Climatic Change* 102 (1–2):225–260. doi:10.1007/s10584-010-9855-0.

Emanuel, Kerry. 2012. *What We Know About Climate Change.* 2nd ed. Cambridge, MA: MIT Press.

FEMA. 2013. New York recovery from Hurricane Sandy: By the numbers. Release Number: NR-210, April 19. http://www.fema.gov/news -release/2013/04/19/new-york-recovery-hurricane-sandy-numbers.

Field, C. B., G. C. Daily, F. W. Davis, S. Gaines, P. A. Matson, J. Melack, and N. L. Miller. 1999. *Confronting Climate Change in California: Ecological Impacts on the Golden State.* Washington, DC: Ecological Society of America.

Food and Agriculture Organization of the United Nations. 2010. Climate-smart agriculture: Policies, practices and financing for food security, adaptation, and mitigation. http://www.fao.org/docrep/013/ i1881e/i1881e00.pdf.

Francis, Jennifer A., and Stephen J. Vavrus. 2012. Evidence linking Arctic amplification to extreme weather in mid-latitudes. *Geophysical Research Letters* 39:L06801.

Frazier, T., N. Wood, B. Yarnal, and D. Bauer. 2010. Influence of potential sea level rise on societal vulnerability to hurricane storm-surge hazards, Sarasota County, Florida. *Applied Geography (Sevenoaks, England)* 30:490–505. doi:10.1016/j.apgeog.2010.05.005.

Gershunov, Alexander, Daniel R. Cayan, and Sam F. Iacobellis. 2009. The great 2006 heat wave over California and Nevada: Signal of an increasing trend. *Journal of Climate* 22:6181–6203. doi:10.1175/2009JCLI2465.1.

Gershunov, A., and K. Guirguis. 2012. California heat waves in the present and future. *Geophysical Research Letters* 39:L18710. doi:10.1029/2012GL052979.

Gillis, Justin. 2013a. Arctic Ice Makes Comeback From Record Low, but Long-Term Decline May Continue. *New York Times.* Septem-

ber 20. http://www.nytimes.com/2013/09/21/science/earth/arctic-ice
-makes-comeback-from-record-low-but-long-term-decline-may-con-
tinue.html?_r=1&&pagewanted=print, accessed September 21, 2013.

Gillis, Justin. 2013b. How high could the tide go? *New York Times.*
January 21. http://www.nytimes.com/2013/01/22/science/earth/see
king-clues-about-sea-level-from-fossil-beaches.html?hp.

Goldenberg, Suzanne. 2013. America's first climate refugees. *The
Guardian,* n.d. http://www.guardian.co.uk/environment/interac-
tive/2013/may/13/newtok-alaska-climate-change-refugees, accessed
June 27, 2013.

Gruber, Nicolas, Claudine Hauri, Zouhair Lachkar, Damian Loher,
Thomas L. Frölicher, Gian-Kasper Plattner. 2012. *Rapid progres-
sion of ocean acidification in the California Current system. Science*
337:220–223.

Hamlet, Alan F., Se-Yeun Lee, Kristian E.B. Mickelson, and Marketa
M. Elsner. 2010. *Climatic Change.* 102 (1–2):103–128.

Hayhoe, Katharine, D. Cayan, C. B. Field, P. C. Frumhoff, E. P. Maurer, N. L. Miller, S. C. Moser, et al. 2004. Emission pathways, climate
change, and impacts on California. *Proceedings of the National Academy of Sciences* 101:12422–12427.

Intergovernmental Panel on Climate Change. 2007. *Climate Change
2007: The Physical Science Basis.* Contribution of Working Group I
to the Fourth Assessment Report of the Intergovernmental Panel on
Climate Change. Edited by S. Solomon, D. Qin, M. Manning, Z. Chen,
M. Marquis, K. B. Averyt, M. Tignor, and H. L. Miller. Cambridge, UK
and New York: Cambridge University Press.

Intergovernmental Panel on Climate Change. 2011. Summary for
Policymakers. In IPCC Special Report on Renewable Energy Sources
and Climate Change Mitigation, edited by O. Edenhofer, R. Pichs-
Madruga, Y. Sokona, K. Seyboth, P. Matschoss, S. Kadner, T. Zwickel,
et al. Cambridge, UK: Cambridge University Press. http://srren.ipcc
-wg3.de/report/IPCC_SRREN_SPM.pdf.

Intergovernmental Panel on Climate Change. 2013. Working Group
I Contribution to the IPCC Fifth Assessment Report Climate Change
2013: The Physical Science Basis Summary for Policymakers. http://
www.climatechange2013.org/spm.

International Biochar Initiative. 2013a. http://www.biochar-interna
tional.org/sustainability.

International Biochar Initiative. 2013b. http://www.biochar-international.org/sites/default/files/March_2013_final.pdf.

Johnstone, J. A., and T. E. Dawson. 2010. Climatic context and ecological implications of summer fog decline in the coast redwood region. *Proceedings of the National Academy of Sciences of the United States of America* 107 (10):4533–4538.

Kaplan, Thomas. 2013. State tells investors that climate change may hurt its finances. *New York Times*. March 26. http://www.nytimes.com/2013/03/27/nyregion/new-york-state-bonds-include-warning-on-climate-change.html?_r=1&&pagewanted=print, accessed June 27, 2013.

Karl, T. R., J. M. Melillo, and T. C. Peterson. 2009. *Global Climate Change Impacts in the United States*. Boston: Cambridge University Press.

Lafferty, Kevin D. 2009. The ecology of climate change and infectious diseases. *Ecology* 90:888–900. doi:10.1890/08-0079.1.

Littell, J. S., E. E. Oneil, D. McKenzie, J. A. Hicke, J. Lutz, and R. A. Norheim. 2010. Forest ecosystems, disturbance, and climatic change in Washington State, USA. *Climatic Change* 102:129–158. doi:10.1007/s10584-010-9858-x.

Loarie, S. R., P. B. Duffy, H. Hamilton, G. P. Asner, C. B. Field, and D. D. Ackerly. 2009. The velocity of climate change. *Nature* 462:1052–1105.

Lobell, D. B., M. B. Burke, C. Tebaldi, M. M. Mastrandrea, W. P. Falcon, and R. L. Naylor. 2008. Prioritizing climate change adaptation needs for food security in 2030. *Science* 319:607–610. doi:10.1126/science.1152339.

Lutz, J. A., J. W. van Wagtendonk, and J. F. Franklin. 2010. Climatic water deficit, tree species ranges, and climate change in Yosemite National Park. *Journal of Biogeography* 37 (5):936–950.

Medellin-Azuara, J., J. J. Harou, M. A. Olivares, K. Madani, J. R. Lund, R. E. Howitt, S. K. Tanaka, et al. 2008. Adaptability and adaptations of California's water supply system to dry climate warming. *Climatic Change* 87 (Supp. 1):S75–S90. doi:10.1007/s10584-007-0355-z.

Meehl, G. A., J. M. Arblaster, and C. Tebaldi. 2005. Understanding future patterns of precipitation extremes in climate model simulations. *Geophysical Research Letters* 32:L18719.

Miller, N., K. Bashford, and E. Strem. 2003. Potential impacts of climate change on California. *Journal of the American Water Resources Association*, Hydrology Paper No. 02035. http://esd.lbl.gov/FILES/about/staff/normanmiller/miller_jawra2003.pdf.

Miller, Norman L., Katharine Hayhoe, Jiming Jin, and Maximilian Auffhammer. 2008. Climate, extreme heat, and electricity demand in California. *Journal of Applied Meteorology and Climatology* 47:1834–1844. doi:10.1175/2007JAMC1480.1.

Mills, L. Scott, Marketa Zimovaa, Jared Oyler, Steven Running, John T. Abatzoglou, and Paul M. Lukacs. 2013. Camouflage mismatch in seasonal coat color due to decreased snow duration. *Proceedings of the National Academy of Sciences of the United States of America* 110 (18):7360–7365. doi:10.1073/pnas.1222724110.

Milly, P. C. D., Julio Betancourt, Malin Falkenmark, Robert M. Hirsch, Zbigniew W. Kundzewicz, Dennis P. Lettenmaier, Ronald J. Stouffer. 2008. Stationarity is dead: Whither water management? *Science* 319 (5863):573–574. doi:10.1126/science.1151915.

Min, S.-K., X. Zhang, F. W. Zwiers, and G. C. Hegerl. 2011. Human contribution to more-intense precipitation extremes. *Nature* 470:378–381. doi:10.1038/nature09763.

Mombelli, Armando. 2012. Major shift in energy policy looms. Swissinfo.ch, Swiss Broadcasting Corporation, November 15. http://www.swissinfo.ch/eng/swiss_news/Major_shift_in_energy_policy_looms.html?cid=33958136.

Mote, P. W., A. F. Hamlet, M. P. Clark, and D. P. Lettenmaier. 2005. Declining mountain snowpack in western North America. *Bulletin of the American Meteorological Society* 86:39–49.

Muller, Richard A. 2013. A pause, not an end, to warming. Op-Ed Contributor. Opinion Pages. *New York Times*. September 25. http://www.nytimes.com/2013/09/26/opinion/a-pause-not-an-end-to-warming.html?_r=0%20accessed%20September%2026,%202013&pagewanted=print, accessed September 27, 2013.

Munich Re Group. 2011. Weather extremes, climate change, Durban 2011. Electronic press folder. Status 25 November. http://www.munichre.com/app_pages/www/@res/pdf/media_relations/press_dossiers/durban_2011/press_folder_durban_2011_en.pdf, accessed September 22, 2013.

NASA. 2013. The current and future consequences of global change. http://climate.nasa.gov/effects/.

National Climate Assessment Development Advisory Committee. 2013. Draft Climate Assessment Report Released for Public Comment v. Jan 2013. http://ncadac.globalchange.gov/.

National Climatic Data Center. 2013. Billion-Dollar Weather/Climate Disasters. NOAA. http://www.ncdc.noaa.gov/billions/.

National Oceanic and Atmospheric Adminstration. 2012. http://oceanexplorer.noaa.gov/explorations/02quest/background/upwelling/upwelling.html.

National Research Council. 2010. Committee on the Development of an Integrated Science Strategy for Ocean Acidification Monitoring, Research, and Impacts Assessment. *Ocean Acidification: A National Strategy to Meet the Challenges of a Changing Ocean.* Washington, DC: The National Academies Press.

National Resources Defense Council. 2009. Ocean acidification: The other CO_2 problem. http://www.nrdc.org/oceans/acidification, accessed September 16, 2013.

National Resources Defense Council. 2013. Climate change health threats in New York. http://www.nrdc.org/health/climate/ny.asp#airpollution, accessed June 27, 2013.

New York State Department of Environmental Conservation. 2010. New York State Climate Action Plan Interim Report. http://www.dec.ny.gov/energy/80930.html, accessed March 29, 2013.

Ocean Carbon and Biogeochemistry Program, Ocean Acidification Subcommittee. 2012. FAQs about ocean acidification: Management and mitigation options. What can we do about ocean acidification on local and regional scales? September 24. http://www.whoi.edu/OCB-OA/page.do?pid=112161#5.

Ortiz, R., K. D. Sayre, B. Govaerts, R. Gupta, G. V. Subbarao, T. Ban, Hodson, D., et al. 2008. Climate change: Can wheat beat the heat? *Agriculture, Ecosystems & Environment* 126:46–58.

Osofsky, Hari M., and Lesley K. McAllister. 2012. *Climate Change Law and Policy.* New York: Wolters Kluwer.

Overland, James E., Jennifer A. Francis, Edward. Hanna, and Muyin. Wang. 2012. The recent shift in early summer Arctic atmospheric circulation. Atmospheric Science. *Geophysical Research Letters* 39 (19), L19804, DOI: 10.1029/2012GL053268.

Parris, A., P. V. Bromirski, D. Burkett, M. Cayan, J. Hall Culver, R. Horton, K. Knuuti, et al. 2012. Global sea level rise scenarios for the US national climate assessment. NOAA Tech Memo OAR CPO-1.

Rahmstorf, Stefan. 2013a. Sea-level rise: Where we stand at the start of 2013. Realclimate.org, January 9. http://www.realclimate.org/in dex.php/archives/2013/01/sea-level-rise-where-we-stand-at-the-start -of-2013.

Rahmstorf, Stefan. 2013b. Sea-level rise: Where we stand at the start of 2013. Part 2. Realclimate.org, January 9. http://www.realclimate .org/index.php/archives/2013/01/sea-level-rise-where-we-stand-at-the-start-of-2013-part-2/.

Rehfeldt, Gerald E.; Nicholas L. Crookston, Marcus V. Warwell, and Jeffrey S. Evans. 2006. Empirical analyses of plant-climate relation-ships for the western United States. *International Journal of Plant Sciences* 167 (6):1123–1150.

Revkin, Andrew. 2013. Climate Panel's Fifth Report Clarifies Hu-manity's Choices. Dot Earth Blog. Opinion Pages. *New York Times*. September 2. http://dotearth.blogs.nytimes.com/2013/09/27/ ipcc-global-warming-report-clarifies-humanitys-choices/?_r=1&& pagewanted=print.

Riordan, B., D. Verbyla, and A. D. McGuire. 2006. Shrinking ponds in subarctic Alaska based on 1950–2002 remotely sensed images. *Journal of Geophysical Research* 111:G04002. doi:10.1029/2005JG000150.

Robock, A., A. Marquardt, B. Kravitz, and G. Stenchikov. 2009. Ben-efits, risks, and costs of stratospheric geoengineering. *Geophysical Research Letters* 36:L19703. doi:10.1029/2009GL039209.

Rosenzweig, C., W. Solecki, A. DeGaetano, M. O'Grady, S. Hassol, and P. Grabhorn, eds. 2011. Responding to Climate Change in New York State: The ClimAID Integrated Assessment for Effective Climate Change Adaptation. Technical Report. Albany, NY: New York State Energy Research and Development Authority.

Roy, Sujoy B., Limin Chen, Evan Girvetz, Edwin P. Maurer, William B. Mills, and Thomas M. Grieb. 2010. Evaluating sustainability of projected water demands under future climate change scenarios. A Tetra Tech, Inc. Report. Prepared for the Natural Resources Defense Council. July. http://timmcgivern.files.wordpress.com/2010/07/tetra_ tech_climate_report_2010_lowres.pdf, accessed September 22, 2013.

Rupp, D. E., P. W. Mote, N. Massey, C. J. Rye, R. Jones, and M. R. Allen. 2012. Did human influence on climate make the 2011 Texas drought more probable? *Bulletin of the American Meteorological Society* 93:1041–1067. doi: 10.1175/BAMS-D-11-00021.1.

Samenow, Jason. 2012. Study: Arctic ice loss may be making North America weather more extreme. Capital Weather Gang Blog. *The Washington Post.* Posted October 10. http://www.washingtonpost .com/blogs/capital-weather-gang/post/study-arctic-ice-loss-mak ing-north-america-weather-more-extreme/2012/10/10/e2f79b88 -1300-11e2-ba83-a7a396e6b2a7_blog.html, accessed September 21, 2013.

Samenow, Jason. 2013. Arctic warming and our extreme weather: no clear link new study finds. Capital Weather Gang Blog. *The Washington Post.* Posted August 19. http://www.washingtonpost.com/ blogs/capital-weather-gang/wp/2013/08/19/arctic-warming-and-our -extreme-weather-no-clear-link-new-study-finds/, accessed September 21, 2013.

Sathaye, J. A., L. L. Dale, P. H. Larsen, G. A. Fitts, K. Koy, S. M. Lewis, and A. Frossard Pereira de Lucena. 2013. Estimating impacts of warming temperatures on California's electricity system. *Global Environmental Change* 23:499–511.

Schwartz, M. D., R. Ahas, and A. Aasa. 2006. Onset of spring starting earlier across the Northern Hemisphere. *Global Change Biology* 12:343–351. doi:10.1111/j.1365-2486.2005.01097.x.

Seager, R., N. Pederson, Y. Kushnir, J. Nakamura, and S. Jurburg. 2012. The 1960s drought and the subsequent shift to a wetter climate in the Catskill Mountains region of the New York City watershed. *Journal of Climate,* 25:6721–6742.

Seager, R., M. Ting, I. Held, Y. Kushnin, J. Lu, G. Vecchi, H.-P. Huang, et al. 2007. Model projections of an imminent transition to a more arid climate in southwestern North America. *Science* 316:1181–1184.

Stec, L., and E. Cordero. 2008. *Cool Cuisine: Taking the Bite out of Global Warming.* Utah: Gibbs Smith.

Stott, P. A., D. A. Stone, and M. R. Allen. 2004. Human contribution to the European heatwave of 2003. *Nature* 432:610–614.

swissinfo.ch. 2013. Energy change is already underway, says Leuthard. September 4. http://www.swissinfo.ch/eng/science_technology/

Energy_change_is_already_underway,_says_Leuthard.html?view
=print&cid=36822514, accessed September 22, 2013.

United Nations Environmental Programme. 2012. 14th Global Meeting of the Regional Seas Conventions and Action Plans Nairobi, Kenya, 1st–3rd October 2012. http://www.unep.org/regionalseas/global meetings/14/RS.14_WP.4.RS.pdf

Warner, Robin, and Clive Schofield, eds. 2012. *Climate Change and the Oceans: Gauging the Legal and Policy Currents in the Asia Pacific and Beyond.* Northampton, MA: Edward Elgar.

Webster, P. J., G. J. Holland, J. A. Curry, and H. R. Chang. 2005. Changes in tropical cyclone number, duration, and intensity in a warming environment. *Science* 309:1844–1846.

Weiss, Kenneth R. 2012. Oceans' rising acidity a threat to shellfish—and humans. *Los Angeles Times.* October 6.

Westerling, A. L., B. P. Bryant, H. K. Preisler, T. P. Holmes, H. Hidalgo, T. Das, and S. Shrestha. 2011a. Climate change and growth scenarios for California wildfire. *Climatic Change* 109 (s1):445–463.

Westerling, A. L., M. G. Turner, E. H. Smithwick, W. H. Romme, and M. G. Ryan. 2011b. Continued warming could transform Greater Yellowstone fire regimes by mid-21st Century. *Proceedings of the National Academy of Sciences of the United States of America* 108 (32):13165–13170.

Wetherald, Richard T., and Syukuro Manabe. 2002. Simulation of hydrologic changes associated with global warming. *Journal of Geophysical Research* 107:4379.

White, M. A., N. S. Diffenbaugh, G. V. Jones, J. S. Pal, and F. Giorgi. 2006. Extreme heat reduces and shifts United States premium wine production in the 21st century. *Proceedings of the National Academy of Sciences of the United States of America* 103:11217–11222.

Wilkinson, R., K. Clarke, M. Goodchild, J. Reichman, and J. Dozier. 2002. *The Potential Consequences of Climate Variability and Change for California: The California Regional Assessment.* Washington, DC: US Global Change Research Program.

Williams, A. P., C. D. Allen, C. Millar, T. Swetnam, J. Michaelsen, C. J. Still, and S. W. Leavitt. 2010. Forest responses to increasing aridity and warmth in southwestern North America. *Proceedings of the National Academy of Sciences of the United States of America* 107 (50):21289–21294.

Williams, J. W., S. T. Jackson, and J. E. Kutzbacht. 2007. Projected distributions of novel and disappearing climates by 2100 AD. *Proceedings of the National Academy of Sciences of the United States of America* 104:5738–5742.

Wootton, J. Timothy, Catherine A. Pfister, and James D. Forester. 2008. Dynamic patterns and ecological impacts of declining ocean pH in a high-resolution multi-year dataset. *Proceedings of the National Academy of Sciences of the United States of America* 105 (48): 18848–18853.

4

The Scientific Consensus on Climate Change: How Do We Know We're Not Wrong?

Naomi Oreskes

In December 2004, *Discover* magazine ran an article on the top science stories of the year. One of these was climate change, and the story was the emergence of a scientific consensus over the reality of global warming. *National Geographic* similarly declared 2004 the year that global warming "got respect" (Roach 2004).

Many scientists felt that respect was overdue. As early as 1995, the Intergovernmental Panel on Climate Change (IPCC) had concluded that "the balance of evidence" supported the conclusion that humans were having an impact on the global climate (Houghton et al. 1995). By 2007, the IPCC's Fourth Assessment Report found a stronger voice, declaring warming "unequivocal" and noting it is "extremely unlikely that the global climate changes of the past fifty years can be explained without invoking human activities" (Alley et al. 2007). Prominent scientists and major scientific organizations have all ratified the IPCC conclusion (Oreskes 2004). Today, all but a tiny handful of climate scientists are convinced that earth's climate is heating up and that human activities are a primary driving cause (Doran and Zimmerman 2009; Anderegg et al. 2010).

Yet, a decade later, Americans continue to wonder. A 2006 poll reported in *Time* magazine found that only just over half

(56 percent) of Americans thought average global temperatures had risen—despite the fact that virtually all climate scientists think they have (The Royal Society 2005). Since 2006, public opinion has wavered—influenced by short-term fluctuations in weather, as well as by political and cultural events whose relationship to climate change is indirect at best (Leiserowitz et al. 2012, and sources cited). But one thing that has remained consistent is a gap between the virtually unanimous opinion of scientists that man-made climate change is underway and the continued doubts of a significant proportion of the American people (Leiserowitz et al. 2012; see also Borick et al. 2011). Moreover, as Jon Krosnick and his colleagues have stressed, while the scientific community has for some time believed the evidence for climate change "justifies substantial public concern," the public has not broadly shared that view (Krosnick et al. 2006; see also Lorenzoni and Pidgeon 2006).

This book addresses the scientific study of climate change and its effects. Its title draws our attention, in particular, to what climate change will mean for our children and grandchildren. By definition predictions are uncertain, and people may wonder why we should spend time, effort, and money addressing a problem that may not affect us for years or decades to come. Some people have gone further, suggesting that it would be foolish to spend time and money addressing a problem that might not actually even exist. After all, how do we really know?

This chapter addresses that issue: how *do* we really know? Put another way, even if there is a scientific consensus, how do we know it's not wrong? If the history of science teaches anything, it is humility. There are numerous historical examples of expert opinion that turned out to be wrong. At the start of the twentieth century, Max Planck was advised not to go into physics because all the important questions had been answered, medical doctors prescribed arsenic for stomach ailments, and geophysicists were confident that continents did not drift. In any scientific community there are individuals who depart from generally accepted views, and occasionally they turn out to be right. At present, there is a scientific consensus that climate change is underway, and that consensus has been stable for more than a decade. But how do we know it's not wrong?

The Scientific Consensus on Climate Change

Let's start with a simple question: what is the scientific consensus on climate change, and how do we know it exists? Scientists do not vote on contested issues, and most scientific questions are far too complex to be answered by a simple yes or no response. So how does anyone know what scientists think about global warming?

Scientists glean their colleagues' conclusions by reading their results in published scientific literature, listening to presentations at scientific conferences, and discussing data and ideas in the hallways of conference centers, university departments, research institutes, and government agencies. For outsiders, this information is difficult to access: scientific papers and conferences are written by experts, for experts, and are difficult for outsiders to understand.

Climate science is a little different. Because of the political importance of the topic, scientists have been motivated and asked to explain their research results in accessible ways, and explicit statements of the state of scientific knowledge are easy to find.

An obvious place to start is the Intergovernmental Panel on Climate Change (IPCC), already discussed in previous chapters. Created in 1988 by the World Meteorological Organization and the United Nations Environment Programme, the IPCC evaluates the state of climate science as a basis for informed policy action, primarily using peer-reviewed and published scientific literature (IPCC 2013a). The IPCC has issued several assessments. In 2001, the IPCC had already stated unequivocally the consensus scientific opinion that earth's climate is being affected by human activities. This view is expressed throughout the report, but perhaps the clearest statement is this: "Human activities ... are modifying the concentration of atmospheric constituents ... that absorb or scatter radiant energy. ... Most of the observed warming over the last 50 years is likely to have been due to the increase in greenhouse gas concentrations" (McCarthy et al. 2001, 21). The 2007 IPCC report amends this to "very likely" (Alley et al. 2007). And the 2013 report added greater specificity, concluding, "It is *extremely likely* [greater than 95 percent confidence] that more than half of the observed increase in global average surface temperature from 1951 to 2010 was caused by the anthropogenic increase in greenhouse gas concentrations and other anthropogenic forcings together" (emphasis in original; IPCC 2013b, SPM-12).

From a historical perspective, the IPCC is a somewhat unusual scientific organization: it was created not to discover new knowledge but to compile and assess existing knowledge on a politically sensitive and economically significant issue. Perhaps its conclusions have been skewed by these extra-scientific concerns, but the IPCC is by no means alone it its conclusions; its results have been repeatedly ratified by other scientific organizations.

All of the major scientific bodies in the United States whose membership's expertise bears directly on the matter have issued reports or statements that confirm the IPCC conclusion. One is the National Academy of Sciences Committee on the

Science of Climate Change report *Climate Change Science: An Analysis of Some Key Questions* (2001), which originated from a White House request. Here is how it opens: "Greenhouse gases are accumulating in Earth's atmosphere as a result of human activities, causing surface air temperatures and subsurface ocean temperatures to rise" (National Academy of Sciences Committee on the Science of Climate Change 2001, 1). The report explicitly addresses whether the IPCC assessment is a fair summary of professional scientific thinking, and answers yes: "The IPCC's conclusion that most of the observed warming of the last 50 years is likely to have been due to the increase in greenhouse gas concentrations accurately reflects the current thinking of the scientific community on this issue" (3).

Other US scientific groups agree. In February 2003, the American Meteorological Society adopted the following statement on climate change: "There is now clear evidence that the mean annual temperature at the Earth's surface, averaged over the entire globe, has been increasing in the past 200 years. There is also clear evidence that the abundance of greenhouse gases has increased over the same period. ... Because human activities are contributing to climate change, we have a collective responsibility to develop and undertake carefully considered response actions" (American Meteorological Society 2003). So too says the American Geophysical Union: "Scientific evidence strongly indicates that natural influences cannot explain the rapid increase in global near-surface temperatures observed during the second half of the 20th century" (American Geophysical Union Council 2003/2007). Likewise the American Association for the Advancement of Science: "The world is warming up. Average temperatures are half a degree centigrade higher than a century ago. ... Pollution from 'greenhouse gases' such as carbon dioxide (CO_2) and methane is at least partly to blame" (Harrison and Pearce 2000). In short, these groups all affirm that global warming is real and substantially attributable to human activities. In 2010, the National Academy of Sciences summarized, "Climate change is occurring, is caused largely by human activities, and poses significant risks for—and in many cases is already affecting—a broad range of human and natural systems" (3).

If we extend our purview beyond the United States, we find this conclusion further reinforced. In 2005, the Royal Society of the United Kingdom, one of the world's oldest and most respected scientific societies, issued a "Guide to Facts and Fictions about Climate Change," debunking various myths asserting that climate change is not occurring, that it is not caused by human activities, that observed changes are within the range of natural variability, that CO_2 is too trivial to matter, that climate models are unreliable, and that the IPCC is biased and does not fairly represent the scientific uncertainties.

The report takes pains to underscore the scientific authority of the IPCC, noting "the IPCC is the world's leading authority on climate change and its impacts" and that its work is backed by the worldwide scientific community.[1] This point was further underscored in 2007, when the National Academies of thirteen countries (the G8 plus another five) issued a joint statement calling attention to the problem of anthropogenic climate change and urging a rapid transition to a low-carbon society (Joint Science Academies 2008).

One website dedicated to evaluating the scientific consensus on climate change counts twenty-seven scientific societies that have formally endorsed the conclusion that "most of the global warming in recent decades can be attributed to human activities"—those just mentioned in North America, Europe, and Australia—as well as thirteen National Academies in Africa.[2]

If we were to do a comprehensive count of scientific societies in Asia, Africa, and South America, the figure would no doubt be still higher.

Consensus reports and statements are drafted through a careful process involving many opportunities for comment, criticism, and revision, so it is unlikely that they would diverge greatly from the opinions of the societies' members. Nevertheless, it could be the case that they downplay dissenting opinions.[3]

One way to test that hypothesis is by analyzing the contents of published scientific papers, which contain the views that are considered sufficiently supported by evidence to merit publication in expert journals. After all, any one can *say* anything, but not anyone can get research results published in a refereed journal.[4] Papers published in scientific journals must pass the scrutiny of critical expert colleagues. They must be supported by sufficient evidence to convince others who know the subject well. So one must turn to the scientific literature to be certain of what scientists really think.

Before the twentieth century, this would have been a trivial task. The number of scientists directly involved in any given debate was usually small. A handful—a dozen, perhaps a hundred, at most—participated, in part because the total number of scientists in the world was very small (Price 1986). Moreover, because professional science was a limited activity, many scientists used language that was accessible to scientists in other disciplines as well as to serious amateurs. It was relatively easy for an educated person in the nineteenth or early twentieth century to read a scientific book or paper and understand what the scientist was trying to say. One did not have to be a scientist to read *The Principles of Geology* or *The Origin of Species*.

Our contemporary world is different. Today, hundreds of thousands of scientists publish over a million scientific papers each year.[5] The American Geophysical Union has 50,000 members in 135 countries, and the American Meteorological Society has 14,000 members. The IPCC reports involved the participation of many hundreds of scientists from scores of countries (Houghton, Jenkins, and Ephraums 1990; Alley et al. 2007). No individual could possibly read all the scientific papers on a subject without making a full-time career of it.

Fortunately, the growth of science has been accompanied by a corresponding growth of tools to manage scientific information. One of the most important of these is the database of the Institute for Scientific Information (ISI). In its Web of Science, the ISI indexes all papers published in refereed scientific journals every year—over 8,500 journals. Using a key word or phrase, one can sample the scientific literature on any subject and get an unbiased view of the state of knowledge.

Figure 4.1 shows the results of an analysis of 928 abstracts, published in refereed journals during the period 1993 to 2003, that I completed in 2004 using the Web of Science database to evaluate the state of scientific debate at that time.[6]

After a first reading to determine appropriate categories of analysis, the papers were divided as follows: (1) those explicitly endorsing the consensus position, (2) those explicitly refuting the consensus position, (3) those discussing methods and techniques for measuring, monitoring, or predicting climate change, (4) those discussing potential, or documenting actual, impacts of climate change, (5) those dealing with paleoclimate change, and (6) those proposing mitigation strategies. How many fell into category 2? That is, how many of these papers presented evidence refuting the statement, "Global climate

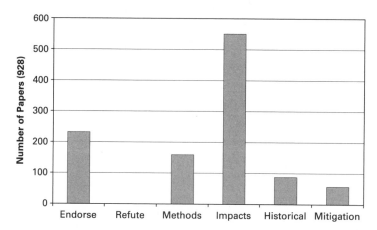

Figure 4.1
A Web of Science analysis of 928 abstracts using the keywords "global climate change." No papers in the sample provided scientific data or theoretical arguments to refute the consensus position on the reality of global climate change.

change is occurring, and human activities are at least part of the reason why"? The answer is remarkable: none.

A few comments are in order. First, it is often challenging to determine exactly what the authors of a paper do think about global climate change. This is a consequence of experts writing for experts: many elements are implicit. If a conclusion is widely accepted, then it is not necessary to reiterate it within the context of expert discussion. Scientists generally focus their discussions on questions that are still disputed or unanswered rather than on matters about which everyone agrees.

This is clearly the case with the largest portion of the papers examined (approximately half of the total)—those dealing with the impacts of climate change. The authors evidently accept the premise that climate change is real and want to track, evaluate, and understand its consequences. Nevertheless, such consequences could, at least in principle, be the results of natural variability rather than human activities. Strikingly, none of the papers used that possibility to argue against the consensus position.

Roughly 15 percent of the papers dealt with methods, and slightly less than 10 percent dealt with paleoclimate change. The most notable trend in the data is the recent increase in such papers; concerns about global climate change have given a boost to research in paleoclimatology and to the development of methods for measuring and evaluating global temperature and climate. Such papers are essentially neutral with respect to the reality of current anthropogenic change: developing better methods and understanding historic climate change are important tools for evaluating current effects, but they do not commit their authors to any particular opinion about those effects. Perhaps some of these authors are in fact skeptical of the current consensus, and this could be a motivation to work on a better understanding of the natural climate variability of the past. But again, none of the papers used that motivation to argue openly against the consensus, and it would be illogical if they did because a skeptical motivation does not constitute scientific evidence. Finally, approximately 20 percent of the papers explicitly endorsed the consensus position, and an additional 5 percent proposed mitigation strategies. In short, by 2003, the basic reality of anthropogenic global climate change was no longer a subject of scientific debate.[7]

Some readers were surprised by this result and questioned the reliability of a study that failed to find arguments against the consensus position when such arguments clearly existed. After all, anyone who watched Fox news or MSNBC or trolled the Internet knew that there was an enormous debate about climate change, right? Well, no.

First, let's make clear what the scientific consensus is. It is over the reality of human-induced climate change. Scientists predicted a long time ago that increasing greenhouse gas emissions could change the climate, and now there is overwhelming evidence that it *is* changing the climate. These changes are *in addition* to natural variability. Therefore, when contrarians try to shift the focus of attention to natural climate variability, they are misrepresenting the situation. No one denies the fact of natural variability, but natural variability alone does not explain what we are now experiencing. Scientists have also documented that many of the changes that are now occurring are deleterious to both human and nonhuman communities (Root et al. 2003; Arctic Council 2004/2005; Hoegh-Guldberg 2005; Parmesan 2006; Adger et al. 2007.) Because of global warming, sea level is rising, humans are losing their homes and hunting grounds, plants and animals are shifting their ranges and in some cases losing their habitats, and extreme weather events (particularly droughts and heat waves) are becoming more common and in some cases more extreme (Kolbert 2006; Flannery 2006; Adger et al. 2007; IPCC 2012).

Second, to say that man-made global warming is underway is not the same as agreeing about what will happen in the future. Much of the continuing debate in the scientific community involves the likely rate of future change. A good analogy is evolution. In the early twentieth century, paleontologist George Gaylord Simpson introduced the concepts of "tempo and mode" to describe questions about the manner of evolution—how fast and in what manner evolution proceeded. Biologists by the mid-twentieth century agreed about the reality of evolution, but there were extensive debates about its tempo and mode. So it is now with climate change. Nearly all professional climate scientists agree that human-induced climate change is underway, but debate continues on tempo and mode.

Third, there is the question of what kind of dissent still exists. My analysis of the published literature was done by sampling published papers, using a keyword phrase that was intended to be fair, accurate, and neutral: "global climate change" (as opposed to, for example, "global warming," which might be viewed as biased). The *total* number of scientific papers published over that ten-year period having anything at all to do with climate change was over ten thousand; it is likely that some of the authors of the unsampled papers expressed skeptical or dissenting views. But given that the sample turned up no dissenting papers at all, professional dissension must have been very limited. Recent work has supported this conclusion, showing that 97–98 percent of professional climate scientists affirm the reality of anthropogenic climate change as outlined by the IPCC (Anderegg et al. 2010; see also Cook et al. 2013). This also affirms the conclusions of Max and Jules Boykoff (2004, see also Freudenburg and Muselli 2010; Boykoff 2011) that the mass media have given air and print space to a handful of dissenters to a degree that is greatly disproportionate with their representation in the scientific community. News articles on climate change, for example, may quote two mainstream scientists and one dissenter, where an accurate reflection of the state of the science would be to quote 30 or 40 mainstream scientists for every dissenter. (On television and radio the situation is even worse, where a debate is set up between one mainstream scientist and one dissenter, as if the actual distribution of views in the scientific community were fifty-fifty.) There are climate scientists who actively do research in the field but disagree with the consensus position, but their number is very, very small. This is not to say that there are not a significant number of *contrarians*, but to point out that the vast majority of them are not climate scientists.

In fact, most contrarians are not even scientists at all. Some, like the physicist Frederick Seitz (who for many years challenged the scientific evidence of the harms of tobacco along with the threat of climate change), were once scientific researchers but not in the field of climate science. (Seitz was a solid-state physicist.) Others, like Michael Crichton, who for many years was a prominent speaker on the contrarian lecture circuit, are novelists, actors, or others with access to the media, but no scientific credentials. What Seitz and Crichton had in common, along with most other contrarians, is that they did little or no new scientific research. They were not producing new evidence or new arguments to be judged by scientists in the halls of science. They were attacking the work of others, and doing so in the court of public opinion and in the mass media.

This latter point is crucial and merits underscoring: the vast majority of books, articles, and websites denying the reality of global warming do not pass the most basic test for what it takes to be counted as scientific—namely, being published in a peer-reviewed journal. Contrarian views have been published in books and pamphlets issued by politically motivated think tanks and widely spread across the Internet (Jacques et al. 2008), but so have views promoting the reality of UFOs or the claim that Lee Harvey Oswald was an agent of the Soviet Union.

Moreover, some contrarian arguments are frankly disingenuous, giving the impression of refuting the scientific consensus when their own data do no such thing. One example will illustrate the point. In 2001, Willie Soon, a physicist at the Harvard-Smithsonian Center for Astrophysics, with several colleagues published a paper entitled "Modeling Climatic Effects of Anthropogenic Carbon Dioxide Emissions: Unknowns and Uncertainties" (Soon et al. 2001). This paper has been widely cited by contrarians as an important example of a legitimate dissenting scientific view published in a peer-reviewed journal.[8] But the issue under discussion is how well models can predict the future—in other words, tempo and mode. The paper does not refute the consensus position, and the authors acknowledge so: "The purpose of [our] review of the deficiencies of climate model physics and the use of GCMs is to illuminate areas for improvement. Our review does not disprove a significant anthropogenic influence on global climate" (Soon et al. 2001, 259; see also Soon et al. 2002).

The authors needed to make this disclaimer because many contrarians do try to create the impression that arguments about tempo and mode undermine the whole picture of global climate change. But they don't. Indeed, one could reject all climate models and still accept the consensus position because models are only one part of the argument—one line of evidence among many.

Is there disagreement over the details of climate change? Yes. Are all the aspects of climate past and present well understood? No, but who has ever claimed that they were? Does climate science tell us what policy to pursue? Definitely not, but it does identify the problem, explain why it matters, and give society insights that can help to frame an efficacious policy response (e.g., Smith 2002; Oreskes, Smith, and Stainforth 2010).

So why does the public have the impression of disagreement among scientists? If the scientific community has forged a consensus, then why do so many Americans have the impression that there is serious scientific uncertainty about climate change?[9]

There are several reasons. First, it is important to distinguish between scientific and political uncertainties. There are reasonable differences of opinion about how best to respond to climate change and even about how serious global warming is relative to other environmental and social issues. Some people have confused—or deliberately conflated—these two issues. Scientists are in agreement about the reality of global climate change, but this does not tell us what to do about it.

Second, climate science involves prediction of future effects, which by definition are uncertain. It is important to distinguish among what is known to be happening now, what is likely to happen based on current scientific understanding, and what might happen in a worst-case scenario. This is not always easy to do, and scientists have not always been effective in making these distinctions. Uncertainties about the future are easily conflated with uncertainties about the current state of scientific knowledge.

Third, scientists have evidently not managed to explain well enough their arguments and evidence beyond their own expert communities. The scientific societies have tried to communicate to the public through their statements and reports on climate change, but what average citizen knows that the American Meteorological Society even exists or visits its home page to look for its climate-change statement?

There is also a deeper problem. Scientists are finely honed specialists trained to create new knowledge, but they generally have limited training in how to communicate to broad audiences and even less in how to defend scientific work against determined and well-financed contrarians (Moser and Dilling 2004, 2007; Hassol 2008; Somerville and Hassol 2011). Moreover, until recently, most scientists have not been particularly anxious to take the time to communicate their message broadly.

Most scientists consider their "real" work to be the production of knowledge, not its dissemination, and often view these two activities as mutually exclusive. Some even sneer at colleagues who communicate to broader audiences, dismissing them as "popularizers" (Olson 2009).

If scientists do jump into the fray on a politically contested issue, they may be accused of "politicizing" the science and compromising their objectivity.[10] This places scientists in a double bind: the demands of objectivity suggest that they should keep aloof from contested issues, but if they don't get involved, no one will know what an objective view of the matter looks like. Scientists' reluctance to present their results to broad audiences has left scientific knowledge open to misrepresentation, and recent events show that there are plenty of people ready and willing to misrepresent it.

It's no secret that politically motivated think tanks such as the American Enterprise Institute and the George C. Marshall Institute have been active for some time in trying to communicate a message that is at odds with the consensus scientific view (Gelbspan 1997, 2004; Mooney 2006; Jacques et al. 2008; Hoggan and Littlemore 2009; Oreskes and Conway 2010). These organizations have successfully garnered a great deal of media attention for the tiny number of scientists who disagree with the mainstream view and for nonscientists, like Crichton, who pronounce loudly on scientific issues.

This message of scientific uncertainty has been reinforced by the public relations campaigns of certain corporations with a large stake in the issue.[11] The most well-known example is ExxonMobil, which in 2000 and 2004 ran highly visible advertising campaigns on the op-ed page of the *New York Times*. Its carefully worded advertisements—written and formatted to look like newspaper columns and called op-ed pieces by

ExxonMobil—suggested that climate science was far too uncertain to warrant action on it.[12] The claims made in these advertisements were not literally untrue, but they were, arguably, very misleading. In 2011 and 2012, ExxonMobil expressed concern about climate change in corporate reports but continued to argue for delay in other venues (Union of Concerned Scientists 2012). Our scientists have long ago concluded that existing research warrants that decisions and policies be made *today*.[13]

In any scientific debate, past or present, one can always find intellectual outliers who diverge from the consensus view. Even after plate tectonics was resoundingly accepted by earth scientists in the late 1960s, a handful of persistent resisters clung to the older views, and some idiosyncratics held to alternative theoretical positions, such as earth expansion. Some of these men were otherwise respected scientists, including Sir Harold Jefferys, one of Britain's leading geophysicists, and Gordon J. F. MacDonald, a one-time science adviser to Presidents Lyndon Johnson and Richard Nixon. Both these men rejected plate tectonics until their dying day, which for MacDonald was in 2002. Does that mean that scientists should reject plate tectonics, that disaster-preparedness campaigns should not use plate-tectonic theory to estimate regional earthquake risk, or that schoolteachers should give equal time in science classrooms to the theory of earth expansion? Of course not. That would be silly and a waste of time. In the case of earthquake preparedness, it would be dangerous as well.

No scientific conclusion can ever be proven, and new evidence may lead scientists to change their views, but it is no more a "belief" to say that earth is heating up than to say that continents move, that germs cause disease, that DNA carries hereditary information, that HIV causes AIDS, and that some synthetic organic chemicals can disrupt endocrine function. You can always find someone, somewhere, to disagree, but these conclusions represent our best current understanding and therefore our best basis for reasoned action (Oreskes 2004).

How Do We Know We're Not Wrong?

Might the consensus on climate change be wrong? Yes, it might be, and if scientific research continues, it is almost certain that some aspects of the current understanding will be modified, perhaps in significant ways. This possibility can't be denied. The relevant question for us as citizens is not whether this scientific consensus *might* be mistaken but rather whether there is any reason to think that it *is* mistaken.

How can outsiders evaluate the robustness of any particular body of scientific knowledge? Many people expect a simple answer to this question. Perhaps they were taught in school that scientists follow "the scientific method" to get correct answers, and they have heard some climate-change deniers suggesting that climate scientists do not follow the scientific method (because they rely on models, rather than laboratory experiments) so their results are suspect. These views are wrong.

Contrary to popular opinion, there is no scientific method (singular). Despite heroic efforts by historians, philosophers, and sociologists, there is no generally agreed-upon answer as to what the methods and standards of science are (or even what they should be). There is no methodological litmus test for scientific reliability and no single method that guarantees valid conclusions that will stand up to all future scrutiny.

A positive way of saying this is that scientists have used a variety of methods and standards to good effect and that philosophers have proposed various helpful criteria for evaluating the methods used by scientists. None is a magic bullet, but each can be useful for thinking about what makes scientific information a reliable basis for action.[14] So we can pose the question: how does current scientific knowledge about climate stand up to these diverse models of scientific reliability?

The Inductive and Deductive Models of Science

The most widely cited models for understanding scientific reasoning are induction and deduction. *Induction* is the process of generalizing from specific examples. If I see 100 swans and they are all white, I might conclude that all swans are white. If I saw 1,000 white swans or 10,000, I would surely think that *all* swans were white, yet a black one might still be lurking somewhere. As David Hume famously put it, even though the sun has risen thousands of times before, we cannot *prove* that it will rise again tomorrow.

Nevertheless, common sense tells us that the sun will rise again tomorrow, even if we can't logically prove that it's so. Common sense similarly tells us that if we had seen ten thousand white swans, then our conclusion that all swans were white would be more robust than if we had seen only ten. Other things being equal, the more we know about a subject, and the longer we have studied it, the more likely our conclusions about it are to be true.

How does climate science stand up to the inductive model? Does climate science rest on a strong inductive base? Yes. Humans have been making temperature records consistently for over 150 years, and nearly all scientists who have looked carefully at these records see an overall temperature increase since the Industrial Revolution (Houghton, Jenkins, and Ephraums 1990; Bruce et al. 1996; Watson et al. 1996; McCarthy et al.

2001; Houghton et al. 2001; Metz et al. 2001; Watson 2001; Weart 2003). According to the Climate Change 2007 Synthesis Report of the IPCC's Fourth Assessment Report, the temperature rise over the 100-year period from 1906 to 2005 was 0.74°C (0.56 to 0.92°C) with a confidence interval of 90 percent (2007a, 27–30). The IPCC's Fifth Assessment Report said temperature for the end of the twenty-first century is "*likely* to exceed 1.5 degrees C relative to 1850 to 1900" for all but one scenario included in the analysis (emphasis in original; IPCC 2013b, SPM-15).

How reliable are the early records? And how do you average data to be representative of the globe as a whole, when most of the early data comes from only a few places, generally in Europe? Scientists have spent quite a bit of time addressing these questions; most have satisfied themselves that the empirical signal is clear (Edwards, 2010). Even if scientists doubted the older records, the more recent data show a strong increase in temperatures over the past thirty to forty years, just when the amount of carbon dioxide and other greenhouses gases in the atmosphere was growing dramatically (McCarthy et al. 2001; Houghton et al. 2001; Metz et al. 2001; Watson 2001). Recently, an independent assessment by the Berkeley Earth Surface Temperature group found that over the past fifty years the land surface warmed by 0.91°C, a result that confirms the prior work by NASA, the National Oceanic and Atmospheric Administration, and the Hadley Centre (Rohde et al. 2013). The Berkeley group has also reviewed the question of the "heat island effect"—the possible exaggeration of the warming effect due to the location of weather stations in urban areas, which are warmer than rural ones because of buildings, concrete, automobiles, and the like—a potential source of error much emphasized by some contrarians (Wickham et al. 2013), and finds that the observed warming cannot be explained away as an artifact of the heat island effect.

The Berkeley study received a good deal of media attention—arguably out of proportion to its scientific significance—because its spokesman, physicist Richard Muller, was previously a self-proclaimed skeptic, and because some of his funding came from the Koch Industries, a Fortune 500 company heavily involved in petroleum refining, oil and gas pipelines, and petrochemicals. (Both Koch brothers are political libertarians who are generally opposed to environmental regulation: David Koch ran in 1980 for Vice President on the Libertarian party ticket, and Charles Koch is one of the founders of the Cato Institute, which has played a large role in US climate change denial; see Oreskes and Conway 2010.) But despite a flurry of media attention, Richard Muller's late-stage conversion had little political, and even less scientific, impact because the conclusions from the instrumental records that he first questioned but then affirmed have been amply corroborated by other independent evidence from tree rings, ice cores, and coral reefs (IPCC 2007b, 438-439). A paper in 2002 by a team led by Jan Esper at the Swiss Federal Research Center, for example, had already demonstrated that tree rings can provide a reliable, long-term record of temperature variability, one that largely (albeit not entirely) agrees with the instrumental records over the past 150 years (Esper, Cook, and Schweingruber 2002).[15]

Muller's reanalysis of existing temperature records raises the fundamental problem facing all inductive science: how many data are enough? If you have counted 10,000 white swans—or 100,000, or even 1,000,000—how do you know that a black swan does not exist elsewhere? And how do you know that the generalization you made from your observations is correct?

After all, other generalizations could also be consistent with your observations.

The logical limitations of the inductive view of science have led some to argue that the core of scientific method is testing theories through logical deductions. *Deduction* is drawing logical inferences from a set of premises—the stock in trade of Sherlock Holmes. In science, deduction is generally presumed to work as part of what has come to be known as the *hypothetico-deductive model*—the model you will find in most textbooks that claim to teach the scientific method (sometimes also called the *deductive-nomological* model, referring to the idea that ultimately science seeks to develop not just hypotheses, but laws).

In this view, scientists develop hypotheses and then test them. Every hypothesis has logical consequences—deductions—and one can try to determine, primarily through experiment and observation, whether the deductions are correct. If they are, they support the hypothesis. If they are not, then the hypothesis must be revised or rejected. It's often considered especially good if the prediction is something that would otherwise be quite unexpected, because that would suggest it didn't just happen by chance.

The most famous example of successful deduction in the history of science is the case of Ignaz Semmelweis, who in the 1840s deduced the importance of hand washing to prevent the spread of infection (Gillispie 1975; Hempel 1965). Semmelweis had noticed that many women were dying of fever after giving birth at his Viennese hospital. Surprisingly, women who had their infants on the way to the hospital—seemingly under more adverse conditions—rarely died of fever. Nor did women who gave birth at another hospital clinic where they were attended by midwives. Not surprisingly, Semmelweis was troubled by this pattern, which seemed to suggest that it was more dangerous to give birth when attended by a doctor than by a midwife, and more dangerous to give birth in a hospital than in a horse-drawn carriage.

In 1847, a friend of Semmelweis, Jakob Kolletschka, cut his finger while doing an autopsy and soon died. Autopsy revealed a pathology very similar to the women who had died after childbirth; something in the cadaver had apparently caused his death. Semmelweis knew that many of the doctors at his clinic routinely went directly from conducting autopsies to attending births, but midwives did not perform autopsies. So he hypothesized that the doctors were carrying cadaveric material on their hands, which was infecting the women (and killed his friend). He deduced that if physicians washed their hands before attending the women, the infection rate would decline. Physicians did, and the infection rate declined, demonstrating the power of the hypothetico-deductive method.

How does climate science stand up to this standard? Have climate scientists made predictions that have come true? Absolutely. The most obvious is the fact of global warming itself. As already noted in previous chapters, scientific concern over the effects of increased atmospheric carbon dioxide is based on physics—the fact that carbon dioxide is a greenhouse gas, a fact that has been known since the mid-nineteenth century. In the early twentieth century, Swedish chemist Svante Arrhenius predicted that increasing carbon dioxide from the burning of fossil fuels would lead to global warming, and by midcentury, a number of other scientists, including G. S. Callendar, Roger Revelle, and Hans Suess, concluded that the effect might soon be noticeable, leading to sea level rise and other global changes (Fleming 1998; Weart 2003). In 1965, Revelle and his colleagues wrote: "By the year 2000, the increase in atmospheric

CO_2 ... may be sufficient to produce measurable and perhaps marked change in climate, and will almost certainly cause significant changes in the temperature and other properties of the stratosphere" (Revelle 1965, 9). This prediction has come true (McCarthy et al. 2001; Houghton et al. 2001; Metz et al. 2001; Watson 2001).

Another prediction fits the category of something unusual that you might not even think of without the relevant theory. In 1980, climatologist Suki Manabe predicted that the effects of global warming would be strongest first in the polar regions. *Polar amplification* was not an induction from observations but a deduction from theoretical principles: the concept of ice albedo feedback. The reflectivity of a material is called its *albedo*. Ice has a high albedo, reflecting sunlight into space much more effectively than grass, dirt, or water. One reason polar regions are as cold as they are is that snow and ice are very effective in reflecting solar radiation back into space. But if the snow starts to melt and bare ground (or water) is exposed, this reflective effect diminishes. Less ice means less reflection, which means more solar heat is absorbed, leading to yet more melting in a feedback loop. So once warming begins, its effects accelerate; Manabe and his colleagues thus predicted that warming would be more pronounced in polar regions than in temperate ones. The Arctic Climate Impact Assessment concluded in 2004 that this prediction had come true (Manabe and Stouffer 1980, 1994; Holland and Bitz 2003; Arctic Council 2004/2005).

Falsification

Ignaz Semmelweis is among the famous figures in the history of science because his work in the 1840s foreshadows the germ theory of disease and the saving of millions of human lives. His story is a great one, told and retold many times. But the story has a twist, because Semmelweis was right for the wrong reason. Cadaveric matter was *not* the cause of the infections: germs were. In later years, this would be demonstrated by James Lister, Robert Koch, and Louis Pasteur, who realized that hand washing was effective not because it removed the cadaveric material, but because it removed the germs associated with that material.

The story illustrates a fundamental flaw with the hypothetico-deductive model—the fallacy of affirming the consequent. If I make a prediction and it comes true, I may assume that my theory is correct. But this would be a mistake, for the accuracy of my deduction does not prove that my hypothesis was correct; my prediction may have come true for other reasons, as indeed Semmelweiss' did. The other reasons may be related to the hypothesis—germs were associated with cadaveric matter—but in other cases the connection may be entirely coincidental. I can convince myself that I have proved my theory right, but this would be self-deception.

This realization led the twentieth-century philosopher Karl Popper to suggest that you can never prove a theory true. Any affirmation of a hypothesis through deduction runs to the risk of the fallacy of affirming the consequent. However, if the prediction does not come true, then you do know that there is something wrong with your hypothesis. Thus Popper emphasized that while science cannot prove a theory true, it can prove it false. Thus, scientific theories must be "falsifiable"—able to be shown, through experiment or observation—that they are false, and the scientific method is useful not to prove theories, but to show them to be false—a view known as *falsificationism* (Popper 1959).

How does climate science hold up to this modification? Can climate models be refuted? Falsification is a bit of a problem for models—not just climate models—because many models are built to forecast the future and the results will not be known for some time. By the time we find out whether the long-term predictions of a model are right or wrong, that knowledge won't be of much use. So while model predictions might be falsifiable in principle, many are not actually falsifiable in practice.

For this reason, many models are tested by seeing if they can accurately reproduce past events—what is sometimes called *retrodiction*. In principle, retrodiction should be a rigorous test: a climate model that fails to reproduce past temperature records is obviously faulty, and could be considered falsified. In reality, it doesn't work quite that way.

Climate models are complex, and they involve many variables—some that are well measured and others that are not. If a model does not reproduce past data very well, most modelers assume that one or more of the model parameters are not quite right, and they make adjustments in an attempt to obtain a better fit. This is generally referred to as *model calibration*, and many modelers consider it an essential part of the process of building a good model. But calibration can make models refutation-proof: the model doesn't get rejected; it gets revised. And given the complexity of climate models, there are myriad ways a model can be revised to ensure that it successfully retrodicts past climate change. Thus, in practice, the idea of falsification is not of great use in judging climate models.

However, one modeler has put his model to the test by making a genuine prediction of the future. When the Philippine volcano Mount Pinatubo erupted in 1991, millions of tons of sulfur dioxide, ash, and dust were thrown into the atmosphere.

Climate modeler James Hansen, then at NASA, realized that these materials were likely to cause a global cooling effect, and that it was possible to use the NASA-Goddard Institute for Space Studies climate model to predict what that cooling would be. The model had been built to simulate long-term global warming, not short-term global cooling, but if the physics of the model were correct, it ought to be able to make this prediction. So Hansen and his team ran the model and forecast a short-term cooling effect of about a half-degree, which would briefly overwhelm the general warming trend from greenhouse gases (Hansen et al. 1992). That prediction came true (Kerr 1993).

This is still only one test, however, and if model results were the only basis for current scientific understanding, there would be grounds for some healthy skepticism. Models are therefore best viewed as heuristic devices: a means to explore what-if scenarios (Oreskes et al. 1994). This is, indeed, how most modelers use them: to answer questions like "If we double the amount of CO_2 in the atmosphere, what is the most likely outcome?"

One way in which modelers address the fact that a model can't be proved right or wrong is to make lots of different models that explore diverse possible outcomes—what modelers call *ensembles*. An example of this is Climateprediction.net, a Web-based mass-participation experiment that enlists members of the public to run climate models on their home computers to explore the range of likely and possible climate outcomes under a variety of plausible conditions.

Over 90,000 participants from over 140 countries have produced tens of thousands of runs of a general circulation model produced by the Hadley Centre for Climate Prediction and Research. Figure 4.2 presents some early results, published in

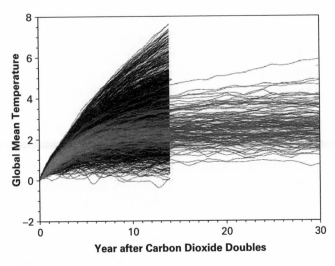

Figure 4.2
Changes in global mean surface temperature (C°) after carbon dioxide values in the atmosphere are doubled. The black lines show the results of 2,579 fifteen-year simulations by members of the general public using their own personal computers. The grey lines show comparable results from 127 thirty-year simulations completed by Hadley Centre scientists on the Met Office's supercomputer (www.metoffice.gov.uk). Figure prepared by Ben Sanderson with help from the climateprediction.net project team.
Source: Reproduced by permission from http://www.climateprediction.net/

the journal *Nature* in 2005, for a steady-state model in which atmospheric carbon dioxide is doubled relative to preindustrial levels and the model earth is allowed to adjust.

The results in black are Climateprediction.net's mass-participation runs; the results in gray come from runs made by professional climate scientists at the Hadley Centre on a supercomputer (Stainforth et al. 2005).

What does an ensemble like this show? For one thing, no matter how many times you run the model, you almost always get the same qualitative result: the earth will warm. The unanswered question is how much and how fast—in other words, tempo and mode.

The models vary quite a bit in their tempo and mode, but nearly all fall within a temperature range of 1° to 8°C (2° to 14°F) within fifteen years after the earth's atmosphere reaches a doubling of atmospheric CO_2 and a large majority fall between 2° and 5°C. Moreover, most of the runs are still warming at that point. The model runs were stopped at year 15 for practicality, but most of them had not yet reached equilibrium: model temperatures were still rising. Look again at figure 4.2. If the general-public model runs had been allowed to continue out to thirty years, as the Hadley Centre scientists' model runs do, many of them would apparently have reached still higher temperatures, perhaps as high as 10–12°C.

How soon will our atmosphere reach a CO_2 level of twice the preindustrial level? The answer depends largely on how much carbon dioxide we humans put into the atmosphere—a parameter that cannot be predicted by a climate model. Note also that in these models CO_2 does not continue to rise: it is fixed at twice preindustrial levels. Nearly all experts now believe that even if major steps are taken soon to reduce the global production of greenhouse gases, atmospheric CO_2 levels will go well above that level. If CO_2 triples or quadruples, then the expected temperature increase will also rise. No one can say precisely when earth's temperature will increase by any specific value, but the models indicate that it almost surely *will* increase. With scant exceptions, the models show the earth warming, and some of them show the earth warming very quickly and substantially.

Is it possible that *all* these model runs are wrong? Yes, because they are variations on a theme. If the basic model conceptualization was wrong in some way, then all the model runs would be wrong, too. Perhaps there is a negative feedback loop that we have not yet recognized. Perhaps the oceans can absorb more CO_2 than we think, or we have missed some other carbon sink (Smith 2002). This is one reason that continued scientific investigation is warranted. But note that Svante Arrhenius and Guy Callendar predicted global warming before anyone ever built a global circulation model (or even had a digital computer). You do not need to have a computer model to predict global warming, and you do not need to have a computer model to know that earth is, currently, warming.

If climate science stands with or without climate models, then is there any information that would show climate science is wrong? Yes. Scientists might discover a mistake in their basic physical understanding that showed they had misconceptualized the whole issue. They could discover that they had overestimated the significance of carbon dioxide and underestimated the significance of some other parameter. But if such mistakes are found, there is no guarantee that correcting them will lead to a more optimistic scenario. It could well be the case that scientists discover neglected factors that show the problem is worse than we'd supposed. (Indeed, some scientists now think this is the case: that we have underestimated the cooling, or "masking" effect of sulfate aerosols, and therefore the impact of greenhouse gases will be worse if or when China, for example, cleans up its air pollution problems.)

Moreover, there is another way to think about this issue. Contrarians have put inordinate amounts of effort into trying to find something that is wrong with climate science, and despite all this effort, they have come up empty-handed. Year after year, the evidence that global warming is real and serious has only strengthened.[16] Perhaps that is the strongest argument of all. Contrarians have repeatedly tried to falsify the consensus, and they have repeatedly failed.

Consilience of Evidence

Most philosophers and historians of science agree that there is no ironclad means by which to prove a scientific theory. But if science does not provide proof, then what is the purpose of induction, hypothesis testing, and falsification? Most would answer that, in various ways, these activities provide a warrant—or a justification—for our views. Do they?

An older view, which has come back into fashion of late, is that scientists look for consilience of evidence. *Consilience* means "coming together," and its use is generally credited to the English philosopher William Whewell, who defined it as the process by which sets of data—independently derived—coincided and came to be understood as explicable by the same theoretical account (Gillispie 1981; Wilson 1998). The idea is not so different from what happens in a legal case. To prove a defendant guilty beyond a reasonable doubt, a prosecutor must present a variety of evidence that holds together in a consistent story. The defense, in contrast, might need to show only that some element of the story is at odds with another to sow reasonable doubt in the minds of the jurors. In other words, scientists are more like lawyers than they might like to admit. They look for independent lines of evidence that hold together.

Do climate scientists have a consilience of evidence? Again the answer is yes. Instrumental records, tree rings, ice cores, borehole data, and coral reefs all point to the same conclusion: things are getting warmer overall. Keith Briffa and Timothy

Osborn of the Climate Research Unit of the University of East Anglia compared the tree-ring analysis by Esper, Cook, and Schweingruber (2002) with six other reconstructions of global temperature between the years 1000 and 2000 (Briffa and Osborn 2002). All seven analyses agree: temperatures increased dramatically in the late twentieth century relative to the record of the previous millennium.

Inference to the Best Explanation

The various problems in trying to develop an account of how and why scientific knowledge is reliable have led some philosophers to conclude that the purpose of science is not proof, but explanation. Not just any explanation will do, however; the best explanation is the one that is consistent with the evidence (e.g., Lipton 1991). Certainly, it is possible that a malicious or mischievous deity placed fossils throughout the geological record to trick us into believing organic evolution—perhaps to test our faith?—but to a scientist this is not the best explanation because it invokes supernatural effects, and the supernatural is beyond the scope of scientific explanation. (It might not be the best explanation to a theologian, either, if that theologian was committed to heavenly benevolence.) Similarly, I might try to explain the drift of the continents through the theory of the expanding earth—as some scientists did in the 1950s—but this would not be the best explanation because it fails to explain why the earth has conspicuous zones of compression as well as tension. The philosopher of science Peter Lipton has put it this way: every set of facts has a diversity of possible explanations, but "we cannot infer something simply because it is a possible explanation. It must somehow be the best of competing explanations" (Lipton 2004, 56). Isaac Newton, in the *Principia Mathematica*, argued that our explanations must invoke causes that we know actually exist—so called *vera causae*. Invoking Martian hunting to explain the extinction of the dinosaurs would not be an inference to the best explanation, because we have no evidence that Martians exist, but invoking a meteorite can be, because large meteorites do.

Best is a term of judgment, so it doesn't entirely solve our problem, but it gets us thinking about what it means for a scientific explanation to be the best available—or even just a good one. It also invites us to ask the question, "Best for what purpose?" For philosophers, *best* generally means that an explanation is consistent with all the available evidence (not just selected portions of it), and that the explanation is consistent with other known laws of nature and other bodies of accepted evidence (and not in conflict with them). In other words, *best* can be judged in terms of the various criterion invoked by *all* the models of science discussed above: Is there an inductive basis? Does the theory pass deductive tests? Do the various elements of the theory fit with each other and with other established scientific information? And is the explanation potentially refutable and not invoking unknown, inexplicable, or supernatural causes?

Contrarians have tried to suggest that the climate effects we are experiencing are simply natural variability. Climate does vary, so this is a *possible* explanation. No one denies that. But is it the *best* explanation for what is happening now? Most climate scientists would say that it's not the best explanation. In fact, it's not even a good explanation—because it is inconsistent with much of what we know.

Should we believe that the global increase in atmospheric carbon dioxide has had a negligible effect, even though basic physics tells us it should be otherwise? Should we believe that the correlation between increased CO_2 and increased temperature is just a peculiar coincidence? If there were no theoretical reason to relate them, and if Arrhenius, Callendar, Suess, and Revelle had not predicted that all this would all happen, then one might well conclude that rising CO_2 and rising temperature were merely coincidental. But we have many reasons to believe that there is a causal connection and no good reason to believe that it is a coincidence. Indeed, the only reason we might think otherwise is to avoid committing to action: if this is just a natural cycle in which humans have played no role, then global warming might go away on its own in due course, and we would not have to spend money or be otherwise inconvenienced to remedy the problem.

And that sums things up. To deny that global warming is real is to deny that humans have become geological agents, changing the most basic physical processes of the earth, and therefore to deny that we bear responsibility for adverse changes that are taking place around us. For centuries, scientists thought that earth processes were so large and powerful that nothing we could do would change them. This was a basic tenet of geological science: that human chronologies were insignificant compared with the vastness of geological time; that human activities were insignificant compared with the force of geological processes. And once they were. But no more. There are now so many of us cutting down so many trees and burning so many billions of tons of fossil fuels that we have become geological agents. We have changed the chemistry of our atmosphere, causing sea level to rise, ice to melt, and climate to change. There is no reason to think otherwise. And, in my view, there is at this point in history no excuse for not taking action to prevent the very significant losses that are likely to ensue—indeed, losses that are already becoming evident.

Notes

1. For additional debunking of myths advanced by climate skeptics, see John Cook et al. (2013), "Skeptical Science: getting skeptical about global warming skepticism." http://www.skepticalscience .com/global-warming-scientific-consensus-intermediate.htm, accessed September 23, 2013.

2. Contrast this with the results of the Intergovernmental Panel on Climate Change's Third and Fourth Assessment Reports, which stated unequivocally that average global temperatures have risen (Houghton et al. 2001; Alley et al. 2007).

3. It should be acknowledged that in any area of human endeavor, leadership may diverge from the views of the led. For example, many Catholic priests endorse the idea that priests should be permitted to marry (Watkin 2004).

4. In recent years, climate-change deniers have increasingly turned to nonscientific literature as a way to promulgate views that are rejected by most scientists (see, for example, Deming 2005).

5. An e-mail inquiry to the Thomson Scientific Customer Technical Help Desk produced this reply: "We index the following number of papers in Science Citation Index—2004, 1,057,061 papers; 2003, 1,111,398 papers."

6. The analysis begins in 1993 because that is the first year for which the database consistently published abstracts. Some abstracts initially compiled were deleted from our analysis because the authors of those papers had put "global climate change" in their key words, but their papers were not actually on the subject.

7. This is consistent with the analysis of historian Spencer Weart, who concluded that scientists achieved consensus in 1995 (see Weart 2003).

8. In e-mails that I received after publishing my essay in *Science* (Oreskes 2004), this paper was frequently invoked. It did appear in the sample.

9. According to *Time* magazine, in 2006 a poll reported that, "64 percent of Americans think scientists disagree with one another about global warming" (*Time* 2006; ABC News/Time/Stanford Poll 2006).

10. Objectivity certainly can be compromised when scientists address charged issues. This is not an abstract concern. It has been demonstrated that scientists who accept research funds from the tobacco industry are much more likely to publish research results that deny or downplay the hazards of smoking than those who get their funds from the National Institutes of Health, the American Cancer Society, or other nonprofit agencies (Bero 2003). On the other hand, there is a large difference between accepting funds from a patron with a clearly vested interest in a particular epistemic outcome and simply trying one's best to communicate the results of one's research clearly and in plain English.

11. Some petroleum companies, such as BP and Shell, have made public efforts to acknowledge the reality of anthropogenic climate change and to refrain from participating in misinformation campaigns (see Browne 1997). Browne began his 1997 lecture by focusing on what he accepted as "two stark facts. The concentration of carbon dioxide in the atmosphere is rising, and the temperature of the Earth's surface is increasing." On the other hand, both BP and Shell were part of the Global Climate Coalition (see Gelbspan 1997, 2004), which promoted disinformation in the early to mid-1990s, during negotiations related to the UN Framework Convention on Climate Change and the Kyoto Protocol. Moreover, after an initial flurry of attention caused by Lord Browne's public statements, BP continued to develop its petroleum resources and only to put modest efforts into developing renewables and carbon sequestration technologies. For an analysis of diverse corporate responses, see Van den Hove et al. (2003).

12. An interesting development in 2003 was that Institutional Shareholders Services advised ExxonMobil shareholders to ask the company to explain its stance on climate-change issues and to divulge financial risks that could be associated with it (see *Planet Ark* 2003).

13. These efforts to generate an aura of uncertainty and disagreement have had an effect. This issue has been studied in detail by academic researchers (see, for example, Boykoff and Boykoff 2004).

14. *Reliable* is a term of judgment. By *reliable basis for action*, I mean that it will not lead us far astray in pursuing our goals, or if it does lead us astray, at least we will be able to look back and say honestly that we did the best we could given what we knew at the time.

15. For further discussion, see Esper, Frank, and Timonen, et al. 2012; and Briffa, Melvin, Osborn et al. 2013.

16. This is evident when the first three IPCC assessments—1990, 1995, and 2001—are compared (Houghton et al. 1990; Bruce et al. 1996; Watson et al. 1996; Houghton et al. 2001; Metz et al. 2001; Watson, 2001; see also Weart 2003).

References

ABC News/Time/Stanford Poll. 2006. March 9–14. http://www.pollingreport.com/enviro3.htm, accessed July 8, 2013.

Adger, Neil, Pramod Aggarwal, Shardul Agrawala, Joseph Alcamo, Abdelkader Allali, Oleg Anisimov, Nigel Arnell, et al. 2007. Climate change impacts, adaptation, and vulnerability: Summary for Policymakers. In *Climate Change 2007: Impacts, Adaptation and Vulnerability. Contribution of Working Group II to the Fourth Assessment Report of the Intergovernmental Panel on Climate Change*, edited by M. L. Parry, O. F. Canziani, J. P. Palutikof, P. J. van der Linden, and C. E. Hanson, 7–22. Cambridge, UK: Cambridge University Press.

Alley, Richard, Terje Berntsen, Nathaniel L. Bindoff, Zhenlin Chen, Amnat Chidthaisong, Pierre Friedlingstein, Jonathan M. Gregory, et al. 2007. IPCC 2007: Summary for Policymakers. In *Climate Change 2007: The Physical Science Basis. Contribution of Working Group I to the Fourth Assessment Report of the Intergovernmental Panel on Climate Change*, edited by S. Solomon, D. Qin, M. Manning, Z. Chen, M. Marquis, K. B. Averyt, M. Tignor, and H. L. Miller. New York: Cambridge University Press.

American Geophysical Union Council. 2003/2007. AGU Position Statement: Human Impacts of Climate. Adopted by Council December 2003. Revised and Reaffirmed December 2007. American Geophysical Union, Washington, DC. http://www.agu.org/sci_pol/positions/climate_change2008.shtml, accessed July 8, 2013.

American Meteorological Society. 2003. Climate change research: Issues for the atmospheric and related sciences. *Bulletin of the American Meteorological Society* 84:508–515. http://www.ametsoc.org/policy/amsstatements_archive.html, accessed July 8, 2013.

Anderegg, W.R.L., James W. Pratt, Jacob Harold, and Stephen H. Schneider. 2010. Expert credibility in climate change. *Proceedings of the National Academy of Sciences* 107 (27):12107–12109.

Arctic Council. 2004/2005. 2004 *Arctic climate impact assessment.* Arctic Council, Oslo, Norway. Cambridge, UK: Cambridge University Press. Available at http://www.amap.no/arctic-climate-impact-assessment-acia, accessed July 8, 2013.

Bero, L. 2003. Implications of the tobacco industry documents for public health and policy. *Annual Review of Public Health* 24:267–288.

Borick, Christopher P., Erick Lachapelle, and Barry Rabe. 2011. Climate compared: Public opinion on climate change in the United States and Canada. Brookings Institute. http://www.brookings.edu/research/papers/2011/04/climate-change-opinion, accessed July 8, 2013.

Boykoff, M. 2011. *Who Speaks for the Climate? Making Sense of Media Coverage of Climate Change.* Cambridge, UK: Cambridge University Press.

Boykoff, M. T., and J. M. Boykoff. 2004. Balance as bias: Global warming and the U.S. prestige press. *Global Environmental Change* 14:125–136.

Briffa, K. R., Thomas M. Melvin, Timothy J. Osborn, Rashit M. Hantemirov, Alexander V. Kirdyanov, Valeriy S. Mazepa, Stepan G. Shiyatov, et al. 2013. Reassessing the evidence for tree-growth and inferred temperature change during the Common Era in Yamalia, northwest Siberia. July 15. *Quaternary Science Reviews* 72:83–107.

Briffa, K. R., and T. J. Osborn. 2002. Blowing hot and cold. *Science* 295:2227–2228.

Browne, E. J. P. 1997. Climate change: The new agenda. Paper presented at Stanford University, May 19, Group Media and Publications, British Petroleum Company.

Bruce, James P., Hoesung Lee, and Erik F. Haites, eds. 1996. *Climate Change 1995: Economic and Social Dimensions of Climate Change. Intergovernmental Panel on Climate Change.* Cambridge, UK: Cambridge University Press.

Cook, J., D. Nuccitelli, S. A. Green, M. Richardson, B. Winkler, R. Painting, R. Way, et al. 2013. Quantifying the consensus on anthropogenic global warming in the scientific literature. *Environmental Research Letters* 8:024024. doi:10.1088/1748–9326/8/2/024024.

Deming, David. 2005. How "consensus" on global warming is used to justify draconian reform. *Investor's Business Daily*, March 18, A16.

Doran, Peter T., and Maggie Kendall Zimmerman. 2009. Examining the scientific consensus on climate change. *Eos* 90 (3):22–23.

Edwards, Paul. 2010. *A Vast Machine: Computer Models, Climate Data, and the Politics of Global Warming.* Cambridge, MA: MIT Press.

Esper, J., E. R. Cook, and F. H. Schweingruber. 2002. Low-frequency signals in long tree-ring chronologies for reconstructing past temperature variability. *Science* 295:2250–2253.

Esper, Jan, David C. Frank, Mauri Timonen, Eduardo Zorita, Rob J. S. Wilson, Jürg Luterbacher, Steffen Holzkämper, et al. 2012. Orbital forcing of tree-ring data. Letters. Published online July 8, 2012. doi: 10.1038. *Nature Climate Change* 1589:1–5.

Flannery, Tim. 2006. *The Weather Makers: How Man Is Changing the Climate and What It Means for Life on Earth.* New York: Atlantic Monthly Press.

Fleming, James Rodger. 1998. *Historical Perspectives on Climate Change.* New York: Oxford University Press.

Freudenburg, William R., and Violetta Muselli. 2010. Global warming estimates, media expectations, and the asymmetry of scientific challenge. *Global Environmental Change* 20 (3):483–491.

Gelbspan, Ross. 1997. *The Heat Is On: The High Stakes Battle over Earth's Threatened Climate.* Reading, MA: Addison-Wesley.

Gelbspan, Ross. 2004. *Boiling Point: How Politicians, Big Oil and Coal, Journalists, and Activists are Fueling the Climate Crisis—And What We Can Do to Avert Disaster.* New York: Basic Books.

Gillispie, Charles C., ed. 1975. Semmelweis. In *Dictionary of Scientific Biography*, vol. 12. New York: Scribner.

Gillispie, Charles C., ed. 1981. *Dictionary of scientific biography.* Vol. 12. New York: Scribner.

Hansen, James, Andrew Lacis, Reto Ruedy, and Makiko Sato. 1992. Potential climate impact of Mount Pinatubo Eruption. *Geophysical Research Letters* 19 (2):215–218.

Harrison, Paul, and Fred Pearce. 2000. *AAAS Atlas of Population and Environment.* Berkeley: University of California Press.

Hassol, Susan Joy. 2008. Improving how scientists communicate about climate change. *Eos* 89 (11):106–107.

Hempel, Carl. 1965. *Aspects of Scientific Explanation, and Other Essays in the Philosophy of Science.* New York: Free Press.

Hoegh-Guldberg, O. 2005. Marine ecosystems and climate change. In *Climate Change and Biodiversity*, edited by T. Lovejoy and L. Hannah, 256–271. New Haven, CT: Yale University Press.

Hoggan, James, and Richard Littlemore. 2009. *Climate Cover-Up: The Crusade to Deny Global Warming*. Vancouver: Greystone Books.

Holland, M. M., and C. M. Bitz. 2003. Polar amplification of climate change in coupled models. *Climate Dynamics* 21:221–232.

Houghton, J. T., Y. Ding, D. J. Griggs, M. Noguer, P. J. van der Linden, X. Dai, K. Maskell, et al., eds. 2001. *Climate Change 2001: The Scientific Basis (Third Assessment Report). Intergovernmental Panel on Climate Change*. Cambridge, UK: Cambridge University Press.

Houghton, J. T., G. J. Jenkins, and J. J. Ephraums, eds. 1990. *Scientific Assessment of Climate Change: Report of Working Group I. Intergovernmental Panel on Climate Change*. Cambridge, UK: Cambridge University Press.

Houghton, J. T., L. G. Meira Filho, B. A. Callander, N. Harris, A. Katteberg, and K. Maskell. 1995. *Climate Change 1995: The Science of Climate Change. Report of Working Group I. Intergovernmental Panel on Climate Change*. Cambridge, UK: Cambridge University Press.

Intergovernmental Panel on Climate Change. 2007a. *Climate Change 2007: Synthesis Report*. Contribution of Working Groups I, II and III to the Fourth Assessment Report of the Intergovernmental Panel on Climate Change [Core Writing Team, Pachauri, R.K and Reisinger, A. (eds.)]. http://www.ipcc.ch/pdf/assessment-report/ar4/syr/ar4_syr.pdf, accessed September 11, 2013.

Intergovernmental Panel on Climate Change. 2007b. *Contribution of Working Group I to the Fourth Assessment Report of the Intergovernmental Panel on Climate Change, 2007*, edited by Solomon, S., D. Qin, M. Manning, Z. Chen, M. Marquis, K. B. Averyt, M. Tignor, and H. L. Miller. Cambridge, UK: Cambridge University Press and New York: Cambridge University Press.

Intergovernmental Panel on Climate Change. 2012. *Managing the Risks of Extreme Events and Disasters to Advance Climate Change Adaptation* edited by Field, C. B., V. Barros, T. F. Stocker, D. Qin, D. J. Dokken, K. L. Ebi, M. D. Mastrandrea, K. J. Mach, G.-K. Plattner, S. K. Allen, M. Tignor, and P. M. Midgley. Cambridge, UK: Cambridge University Press.

Intergovernmental Panel on Climate Change. 2013a. Organization. http://www.ipcc.ch/organization/organization.shtml, accessed September 15, 2013.

Intergovernmental Panel on Climate Change. 2013b. Working Group I Contribution to the IPCC Fifth Assessment Report Climate Change 2013: The Physical Science Basis Summary for Policymakers. http://www.climatechange2013.org/images/uploads/WGIAR5-SPM_ Approved27Sep2013.pdf.

Jacques, Peter J., Riley E. Dunlap, and Mark Freeman. 2008. The organisation of denial: Conservative think tanks and environmental scepticism. *Environmental Politics* 17 (3):349–385.

Joint Science Academies. 2008. Joint Science Academies' Statement: Climate Change Adaptation and the Transition to a Low Carbon Society. Academies of Science for the G8+5 countries. http://www.science.org.au/policy/climatechange-g8+5.pdf, accessed September 23, 2013.

Kerr, Richard. 1993. Pinatubo global cooling on target. *Science* 259:5095.

Kolbert, Elizabeth. 2006. *Field Notes from a Catastrophe*. New York: Bloomsbury.

Krosnick, Jon A., Allyson I. Holbrook, Laura Lowe, Penny S. Visser. 2006. The origins and consequences of democratic citizens' policy agendas: A study of popular concern about global warming. *Climate Change* 77 (1–2):7–43.

Leiserowitz, A., E. Maibach, C. Roser-Renouf, and J. Hmielowski. 2012. *Global Warming's Six Americas in March 2012 & Nov. 2011.* New Haven, CT: Yale Project on Climate Change Communication. http://environment.yale.edu/climate/files/Six-Americas-March-2012 .pdf, accessed July 8, 2013.

Lipton, Peter. 1991. *Inference to the Best Explanation*. Oxford: Routledge.

Lipton, Peter. 2004. *Inference to the Best Explanation*. 2nd ed. Oxford: Routledge.

Lorenzoni, Irene, and Nick F. Pidgeon. 2006. Public views on climate change: European and USA perspectives. *Climatic Change* 77 (1–2):73–95.

Manabe, S., and R. J. Stouffer. 1980. Sensitivity of a global climate model to an increase of CO_2 concentration in the atmosphere. *Journal of Geophysical Research* 85 (C10):5529–5554.

Manabe, S., and R. J. Stouffer. 1994. Multiple-century response of a coupled ocean-atmosphere model to an increase of atmospheric carbon dioxide. *Journal of Climate* 7 (1):5–23.

McCarthy, James J., Osvaldo F. Canziani, Neil A. Leary, David J. Dokken, and Kasey S. White, eds. 2001. *Climate Change 2001: Impacts, Adaptation and Vulnerability. Intergovernmental Panel on Climate Change.* Cambridge, UK: Cambridge University Press.

Metz, Bert, Ogunlade Davidson, Rob Swart, and Jiahua Pan, eds. 2001. *Climate Change 2001: Mitigation. Intergovernmental Panel on Climate Change.* Cambridge, UK: Cambridge University Press.

Mooney, Chris. 2006. *The Republican War on Science.* New York: Basic Books.

Moser, S. C., and L. Dilling. 2004. Making climate hot: Communicating the urgency and challenge of global climate change. *Environment* 10:32–46.

Moser, S. C., and L. Dilling, eds. 2007. *Creating a Climate for Change: Communicating Climate Change and Facilitating Social Change.* Cambridge, UK: Cambridge University Press.

National Academy of Sciences. 2010. *Advancing the Science of Climate Change.* By America's Climate Choices: Panel on Advancing the Science of Climate Change; National Research Council. Washington, DC: National Academies Press.

National Academy of Sciences Committee on the Science of Climate Change. 2001. *Climate Change Science: An Analysis of Some Key Questions.* Washington, DC: National Academy Press.

Olson, Randy. 2009. *Don't Be Such a Scientist: Talking Substance in an Age of Style.* Washington, D.C.: Island Press.

Oreskes, N. 2004. Beyond the ivory tower: The scientific consensus on climate change. *Science* 306 (5702):1686.

Oreskes, Naomi, and Erik M. Conway. 2010. *Merchants of Doubt: How a Handful of Scientists Obscured the Truth on Issues from Tobacco Smoke to Global Warming.* New York: Bloomsbury.

Oreskes, Naomi, Kristin Shrader-Frechette, and Kenneth Belitz. 1994. Verification, validation, and confirmation of numerical models in the earth sciences. *Science* 263:641–646.

Oreskes, Naomi, Leonard Smith, and David Stainforth. 2010. Adaptation to global warming: Do climate models tell us what we need to know? *Philosophy of Science* 77:1012–1028.

Parmesan, Camille. 2006. Ecological and evolutionary responses to recent climate change. *Annual Review of Ecology, Evolution, and Systematics* 37:637–669.

Planet Ark. 2003. ISS in favor of ExxonMobil global warming proposals. May 19. http://www.planetark.com/dailynewsstory.cfm/newsid/20824/story.htm, accessed July 8, 2013.

Popper, K. R. 1959. *The Logic of Scientific Discovery.* London: Hutchinson.

Price, Derek de Solla. 1986. *Little Science, Big Science—And Beyond.* New York: Columbia University Press.

Revelle, Roger. 1965. Atmospheric carbon dioxide. In *Restoring the Quality of Our Environment: A Report of the Environmental Pollution Panel, 111–33.* Washington, DC: President's Science Advisory Committee.

Roach, John. 2004. The year global warming got respect. *National Geographic.* December 29. http://news.nationalgeographic.com/news/2004/12/1229_041229_climate_change_consensus.html accessed July 8, 2013.

Rohde, R., R. A. Muller, R. Jacobsen, E. Muller, S. Perlmutter, A. Rosenfeld, J. Wurtele, et al. 2013. A new estimate of the average earth surface land temperature spanning 1753 to 2011. *Geoinformatics & Geostatistics: An Overview* 1:1. doi:10.4172/gigs.1000101.

Root, T. L., J. T. Price, K. R. Hall, S. H. Schneider, C. Rosenzweig, and J. A. Pounds. 2003. Fingerprints of global warming on wild animals and plants. *Nature* 421:57–60.

The Royal Society. 2005. *A guide to facts and fictions about climate change.* March. http://royalsociety.org/uploadedFiles/Royal_Society_Content/News_and_Issues/Science_Issues/Climate_change/climate_facts_and_fictions.pdf, accessed September 16, 2013.

Smith, Leonard A. 2002. What might we learn from climate forecasts? *Proceedings of the National Academy of Sciences of the United States of America* 99 (Suppl. 1):2487–2492.

Somerville, Richard C., and Susan Joy Hassol. 2011. Communicating the science of climate change. *Physics Today* 64:48–53.

Soon, W., S. Baliunas, S. B. Idso, K. Y. Kondratyev, and E. S. Posmentier. 2001. Modeling climatic effects of anthropogenic carbon dioxide emissions: Unknowns and uncertainties. *Climate Research* 18:259–275.

Soon, W., S. Baliunas, S. B. Idso, K. Y. Kondratyev, and E. S. Posmentier. 2002. Modeling climatic effects of anthropogenic carbon dioxide emissions: Unknowns and uncertainties, reply to Risbey. *Climate Research* 22:187–188.

Stainforth, D., T. Aina, C. Christensen, M. Collins, N. Faull, D. J. Frame, J. A. Kettleborough, et al. 2005. Uncertainty in predictions of the climate response to rising levels of greenhouse gases. *Nature* 433:403–406.

Time. 2006. Americans see a climate problem. March 26.

Union of Concerned Scientists. 2012. *A Climate of Corporate Control: Company Profiles.* http://www.ucsusa.org/scientific_integrity/abuses_of_science/corporate-climate-company-profiles.html.

Van den Hove, Sybille, Marc Le Menestrel, and Henri-Claude de Bettignies. 2003. The oil industry and climate change: Strategies and ethical dilemmas. *Climate Policy* 2:3–18.

Watkin, Daniel J. 2004. Roman Catholic priests' group calls for allowing married clergy members. *New York Times.* April 28, B5.

Watson, Robert T., ed. 2001. *Climate Change 2001: Synthesis Report. Intergovernmental Panel on Climate Change.* Cambridge, UK: Cambridge University Press.

Watson, R. T., Marufu C. Zinyowera, and Richard H. Moss, eds. 1996. *Climate Change 1995: Impacts, Adaptations and Mitigation of Climate Change—Scientific-Technical Analyses. Intergovernmental Panel on Climate Change.* Cambridge, UK: Cambridge University Press.

Weart, Spencer R. 2003. *The Discovery of Global Warming.* Cambridge, MA: Harvard University Press.

Wickham, C., R. Rohde, R. A. Muller, J. Wurtele, J. Curry, D. Groom, R. Jacobsen, et al. 2013. Influence of urban heating on the global temperature land average using rural sites identified from MODIS classifications. *Geoinformatics & Geostatistics: An Overview* 1:2. doi:10.4172/gigs.1000104.

Wilson, Edward 0. 1998. *Consilience: The Unity of Knowledge.* New York: Alfred A. Knopf.

5

Climate Change: How the World Is Responding

Joseph F. C. DiMento and Pamela Doughman

Scientists warn that it is too late to avoid climate change, but we can act now to ease the problem for our children and grandchildren. A large part of the United States population has come to this understanding. A 2006 *Los Angeles Times*/Bloomberg survey found that almost half of Americans think global warming is caused more by human activities than by natural changes in the climate and that 56 percent believe the government could do more to address the problem (Boxall 2006). A poll conducted by Stanford University in 2013 found that more than 80 percent of adults in the United States favor preparing for climate change through measures such as stronger building codes and preventing new construction in vulnerable coastal areas (*USA Today* 2013). They are basing their opinions on the evidence presented by an army of climate scientists and an almost equally large group of policy analysts, politicians, and other government leaders.

We can ease the impact of climate change by mitigating greenhouse gas (GHG) emissions and taking steps to adapt to climate change effects. Efforts to mitigate and adapt to climate change are underway at the international, regional, national, and local levels.

Recognition that meaningful actions to protect our climate are possible has come in different forms across the globe, through modest binding commitments by the developed world to reduce GHG emissions, voluntary efforts, and efforts limited to the major sources of emissions. Most countries have concluded that it makes sense to meet regularly to review current science, policies, and programs to decrease GHG emissions and increase resiliency to climate change. Perhaps meetings are not the most efficient way to address a global environmental challenge, but they are one of the few options that countries have to face the problem, because greenhouse emissions in one place impact climate change globally. A lack of resiliency to the effects of climate change can have international ramifications as well.

Nonetheless, there is no need to rely solely on international initiatives to reduce the impacts of climate change. National, subnational, and local GHG emission reductions can lead the way for others to take action while reducing pollution, improving infrastructure, and promoting economic growth closer to home.

Similarly, reducing vulnerability to climate change can start with local and state governments. Disaster planning and preparedness, infrastructure reinforcement and upgrades, and changes in land use planning can reduce the costs of extreme weather events where they are expected to become more frequent with climate change. Information on expected changes in precipitation patterns can be used to modify agricultural practices, forestry management, and water management to adapt to new conditions. Fisheries management and cultivation of seagrass beds can help reduce losses from changing ocean acidity levels by utilizing information on patterns and variability of local acidity indices. International cooperation improves the availability of information on changes expected from climate change and best practices for adaptation.

In their official positions, countries have moved with varying senses of urgency. Some are fearful that their land mass and resource base are threatened in the near future. Others have concluded that economic growth and other environmental challenges are more important than climate change. Several nations have gone beyond the official international requirements and have taken steps to make significant reductions in their contributions to climate change. The link between climate policy and other national policies (such as energy independence, technological leadership, and other green goals) has been recognized, so climate-change policy might achieve several important national objectives. States and provinces have recognized the benefits of some climate-change policies, as have utilities, high-tech companies, and even some energy giants.

These conclusions have come in fits and starts, with significant shifts in the positions of important countries, including the United States, and disagreements within countries and within states and businesses about how many changes they should make. The costs of change are important, and internationally they have been calculated and balanced against the benefits of reducing emissions of greenhouse gases and against the benefits of moving slowly in different ways. Some calls for mitigation of climate change focus on climate justice and climate refugees (Garnier and Faimali 2010, 175), while practitioners in the field call for focusing on developmental priorities in the design and implementation of climate-adaptation projects (Anguelovski and Roberts 2011). The understanding that major shifts in climate-change policy will produce winners and losers has set the political stage for world opinion.

Independently of official international and national responses, smaller organizations and even individuals have turned into action their beliefs that climate change can be mitigated. Those actions reflect very different understandings of how important global warming is compared to other social goals. In a 2013 national survey, for example, more than six in ten said corporations, industry, and citizens should be doing more to address global warming and more than half said Congress and the president should be doing more. More than 60 percent of Americans supported the following actions: "Providing tax rebates for people who purchase energy-efficient vehicles or solar panels"; "Funding more research into renewable energy sources"; "Regulating CO_2 as a pollutant"; and "Requiring fossil fuel companies to pay a carbon tax and using the money to pay down the national debt" (Leiserowitz et al. 2013).

In this chapter, we describe how the world has reacted to the scientific message. We focus on the supporters and opponents of international efforts to control GHG emissions through international laws—why there have been such different understandings of what (if anything) needs to be done, what US climate-change policy has been, and the responses of state and local governments and industry.

Efforts at International Cooperation for GHG Reduction and a Legal Response

In the face of mounting evidence that action was needed to reduce the growth of greenhouse gases in the atmosphere, 10,000 delegates from around the world met in Kyoto, Japan, in 1997 to make good on commitments made in the United Nations Framework Convention on Climate Change (UN-FCCC) in 1992. After a series of late-night negotiations, the delegates produced an international agreement—the Kyoto Protocol—that called on industrialized countries as a group to reduce emissions of greenhouse gases during the 2008–2012 commitment period to at least 5 percent below 1990 levels. Nations with reduction commitments had different targets ranging from –8 percent from the base period to +10 percent for the first commitment period.

Some thought the Kyoto Protocol was a promising start to a difficult problem; others wavered. The United States thought it was a promising start, then wavered, then reneged. Russia hesitated, but its ultimate support for the Kyoto Protocol allowed the agreement to go into effect in 2005. In twenty-five years of international efforts to address climate change, the 1997 Kyoto Protocol marked the first agreement on binding limits of GHG emissions.

Pre-Kyoto

Before the 1997 Kyoto Protocol, the dominant focus of international environmental policy was on cooperative research and voluntary goals for reducing emissions. In 1979, the First World Climate Conference called for greater cooperation to study climate change and prevent it from getting worse (Zillman 2009).

The late 1980s were a period of growing awareness of global ecological problems and their social, economic, and health dilemmas. Scientists discovered the weakening of the earth's protective ozone layer. An international agreement, the Montreal Protocol, was signed in 1987 to control emissions of ozone-depleting substances. The publication of *Our Common Future* by the World Commission on Environment and Development (Bruntland 1987) focused international attention on the need for sustainable development and on the links among the environment, society, and the economy. The book emphasized that damaging any one of these three elements weakens the other two—either in the current generation or in those that follow. Awareness of climate change was also heightened during this time period, through publication of the First Assessment Report of the Intergovernmental Panel on Climate Change, or IPCC (IPCC 1990). This report said that human activities are substantially increasing the amount of GHGs in the atmosphere. Although effects vary according to region and time period, the report predicted that the average temperature of the earth's surface will become warmer than the average earth surface temperatures known during the past 10,000 years.

This period's sense of urgency was reflected in the frequency and magnitude of the international environmental negotiations on climate change that were conducted during this time (see table 5.1). Although eleven years passed between the First World Climate Conference in 1979 and the Second World Climate Conference in 1990, a flurry of negotiations led to the 1992 United Nations Conference on Environment and Development in Rio de Janeiro, Brazil. Several international environmental agreements were signed at the Rio Conference, including the Framework Convention on Climate Change (FCCC). The FCCC set out general parameters and principles for international efforts to address climate change; details were to be worked out in subsequent negotiations. The FCCC aims for

stabilization of greenhouse gas concentrations in the atmosphere at a level that would prevent dangerous anthropogenic interference with the climate system. Such a level should be achieved within a time-frame sufficient to allow ecosystems to adapt naturally to climate change, to ensure that food production is not threatened and to enable economic development to proceed in a sustainable manner. (United Nations 1992, art. 2)

In support of the idea that climate change is a common concern and the world's nations have different responsibilities to the earth's environment, developed countries and developing countries have different commitments under the FCCC. All signers agreed to provide information and work together, but developed countries agreed to be leaders in addressing climate change. They decided to reduce emissions of carbon dioxide and other greenhouse gases not controlled by the Montreal Protocol to 1990 levels by 2000 through the use of appropriate and cost-effective technologies and means. Reductions can be achieved individually or jointly. These commitments were not asked of developing countries. Instead, the FCCC encourages developing countries to reduce GHG emissions (without a legally binding target) and promote sustainable development. The FCCC also provides an incentive for developing countries: developed countries agreed to fund developing countries' compliance with the treaty.

The FCCC recognizes that an important part of sustainable development is the conservation and enhancement of *sinks* (which remove greenhouse gases and aerosols from the atmosphere) and reservoirs of greenhouse gases. According to the agreement, sinks are "biomass, forests and oceans as well as other terrestrial, coastal and marine ecosystems."

GHG emissions were to be reduced based on currently available scientific knowledge, despite areas of continuing uncertainty:

The Parties should take precautionary measures to anticipate, prevent or minimize the causes of climate change and mitigate its adverse effects. Where there are threats of serious or irreversible damage, lack of full scientific certainty should not be used as a reason

Table 5.1
Major International Climate Meetings and Conferences

1979	First World Climate Conference
1988	World Meteorological Association and United Nations Environmental Program establish the Intergovernmental Panel on Climate Change
1990	Second World Climate Conference
1990	UN General Assembly establishes Intergovernmental Negotiating Committee for the Framework Convention on Climate Change
1992	Signing of Framework Convention on Climate Change at the United Nations Conference on Environment and Development. It provides: *Developed countries:* Reduce greenhouse gases to 1990 levels; Provide cost of compliance of developing countries. *Voluntary compliance encouraged for developing countries.* Cost effectiveness and sinks recognized.
1994	Framework Convention on Climate Change enters into force
1995	First Conference of the Parties, Berlin, Germany
1996	Second Conference of the Parties, Geneva, Switzerland
1997	Third Conference of the Parties, Kyoto, Japan
1997	Kyoto Protocol signed. Aims to reduce overall emissions by at least 5 percent below 1990 levels in the period 2008–2012
1998	Fourth Conference of the Parties, Buenos Aires, Argentina
1999	Fifth Conference of the Parties, Bonn, Germany
2000	Sixth Conference of the Parties, Part I, The Hague, Netherlands
2001	Sixth Conference of the Parties, Part II, Bonn, Germany
2001	Seventh Conference of the Parties, Marrakech, Morocco
2002	Eighth Conference of the Parties, New Delhi, India
2003	Ninth Conference of the Parties, Milan, Italy
2004	Tenth Conference of the Parties, Buenos Aires, Argentina

2005	Kyoto Protocol goes into effect in February
2005	First Meeting of the Parties to the Kyoto Protocol and Eleventh Conference of the Parties, Montreal, Canada
2006	Second Meeting of the Parties to the Kyoto Protocol and Twelfth Conference of the Parties, Nairobi, Kenya
2007	Third Meeting of the Parties to the Kyoto Protocol and Thirteenth Conference of the Parties, Bali, Indonesia IPCC releases Fourth Assessment Report, declares world is warming largely due to human activities
2008	Fourth Meeting of the Parties to the Kyoto Protocol and Fourteenth Conference of the Parties, Poznan, Poland
2009	Fifth Meeting of the Parties to the Kyoto Protocol and Fifteenth Conference of the Parties, Copenhagen, Denmark
2010	Sixth Meeting of the Parties to the Kyoto Protocol and Sixteenth Conference of the Parties, Cancun, Mexico
2011	Seventh Meeting of the Parties to the Kyoto Protocol and Seventeenth Conference of the Parties, Durban, South Africa
2012	Eighth Meeting of the Parties to the Kyoto Protocol and Eighteenth Conference of the Parties, Doha, Qatar
2013	Ninth session of the Conference of the Parties serving as the meeting of the Parties to the Kyoto Protocol and Nineteenth Conference of the Parties, Warsaw, Poland

Source: UNFCCC 2013c. Meetings. http://unfccc.int/meetings/
items/6240.php

for postponing such measures, taking into account that policies and measures to deal with climate change should be cost-effective so as to ensure global benefits at the lowest possible cost. (United Nations 1992, art. 3)

FCCC nations—those that ratify, approve, accede to, or accept the agreement—promised to meet each year at a conference of the parties (COP) to discuss advances in scientific understanding of climate change and update details regarding implementation. At these conferences, thousands of delegates and observers address the almost innumerable steps required to implement the agreement. Opened for signature in 1992, the FCCC went into effect in 1994. As of June 2013, FCCC participants now include 195 parties and three observer states. (UNFCCC 2013d). This level of support for international efforts to address climate change has not been seen since.

The first of the FCCC conferences was held in Berlin in 1995, and negotiations were tense. With significant leadership from India, however, a majority of the parties agreed to establish binding reduction targets for developed countries but not for developing countries. At the time, the United States supported the agreement (Carpenter, Chasek, and Wise 1995).

Small island states, the Netherlands, Germany, and Malaysia, among others, argued for stronger reductions in GHG emissions (Carpenter et al. 1995). In contrast, some oil-exporting countries said the conference had gone too far. Russia and the United Arab Emirates questioned the scientific basis for action, and Venezuela challenged the conclusion that existing commitments were inadequate. Saudi Arabia, Venezuela, and Kuwait had concerns about the decisions reached at the first conference (Carpenter et al. 1995).

By the time the parties met for the second formal FCCC conference in 1996, significant disagreements remained, even though smaller meetings had been held between the two conferences. One of the central points of conflict was whether to use the Second Assessment Report of the IPCC (1995) as the basis for negotiating binding agreements on developed countries to reduce GHG emissions. Opponents—mainly oil-exporting countries—disagreed with the report's conclusion that "the balance of evidence" indicates human-caused emissions of greenhouse gases are having a "discernible" influence on global climate change. Some environmentalists were encouraged to see the US insurance industry support early and substantial reductions of GHG emissions to help reduce the risk of severe weather and associated property damage (Carpenter et al. 1996).

A number of other issues were also controversial. The United States originally opposed legally binding targets, but decided to support them provided that a country could reach its targets by acquiring a right of emission from another entity that has met or does not have to meet a target. Another issue was that developing countries and oil-exporting countries wanted stronger protections against possible economic losses. A number of these nations argued against market-based approaches (such as tradable carbon dioxide permits) because the poor, with perhaps the greatest demand for CO_2 reduction, would be least able to pay to communicate what they need to the market.

Despite the turmoil surrounding the first two FCCC conferences of the parties, several concepts were developed that would be part of the basis for negotiations of the Kyoto Protocol at the third conference of the parties in 1997 (UNFCCC Conference of the Parties 1996):

Legally binding reductions in GHG emissions should be established for industrialized countries;

Developing countries should be exempt;

The Second Assessment Report should "provide a scientific basis for urgently strengthening action ... to limit and reduce emissions of greenhouse gases" (71); and

The IPCC needs to further reduce scientific uncertainties, "in particular regarding socio-economic and environmental impacts on developing countries, including those vulnerable to drought, desertification or sea-level rise" (71).

Kyoto

Some countries maintained strong concerns over FCCC implementation. Many governments believe current measures are inadequate, and some argue they cannot afford to do more. These positions were raised again during the negotiations at Kyoto in 1997. Nonetheless, the Kyoto Protocol is consistent with the earlier agreements among the FCCC countries.

Specifically, in the twenty-seven articles of the Kyoto Protocol, Annex I countries agreed to reduce GHG emissions by "assigned amounts" specific to each country: "The parties included in Annex I shall, individually or jointly, ensure that their aggregate anthropogenic carbon dioxide equivalent emissions ... do not exceed their assigned amounts ... with a view to reducing their overall emissions of such gases by at least 5 percent below 1990 levels in the commitment period 2008 to 2012" (article 3.1). Annex I countries include most industrialized and some central European nations. Their reduction commitment ranged from 92 percent (change from the base year) to 108 percent.

No developing country signing the FCCC committed to any assigned amount or quantitative limit on GHG emissions. The role of developing countries in reducing GHG emissions is not specified in the Kyoto Protocol other than as potential partners in efforts by Annex I countries to meet their commitments (articles 4 and 6) and as recipients of technology transfer (article 3.14). Developing countries are mentioned as potentially subject to undesirable side effects that may result from reduction of greenhouse gases. To guard against such outcomes, article 2.3 of the Kyoto Protocol requires annex I countries to "strive to implement policies and measures under this Article in such a way as to minimize adverse effects, including the adverse effects of climate change, effects on international trade, and social, environmental and economic impacts on other Parties, especially developing country Parties." Similarly, article 3.14 of the protocol requires annex I countries to "strive to implement the commitments mentioned in paragraph I above in such a way as to minimize adverse social, environmental and economic impacts on developing country Parties."

The fifth conference of the parties in 1999 addressed details of emission trading, the Clean Development Mechanism, joint implementation (the so-called flexibility mechanisms), accounting of GHG emissions, and development of a credible compliance system. Emissions trading occurs among industrialized nations. Joint implementation offers emission reduction units for financing projects in other developed countries (such as power plant conversions). The Clean Development Mechanism provides credit (certified emissions reductions) for financing emissions-reducing or emission-avoiding projects in developing countries.

Kyoto Signatories

Countries representing 55 percent of 1990 emissions from the industrialized countries (the Annex 1 countries) had to ratify, approve, accede to, or accept the Kyoto Protocol to bring it into effect. On the basis of those countries that had already committed, this meant either the United States or Russia had to accept the agreement. Ninety days after Russia ratified it (on November 18, 2004), the Kyoto Protocol went into effect on February 16, 2005. By then, several additional conferences of the parties had been held where compromises were reached, issues set aside for later consideration, and important programs developed for eventual implementation of the protocol. The first meeting of the parties of the Kyoto Protocol took place simultaneously with a conference of the parties of the FCCC in December 2005 in Montreal, and the parties met (for the second meeting of the parties and twelfth conference of the parties) again in Nairobi in November 2006.

By 2005, however, many of the 124 countries that had already ratified, accepted, approved, or acceded to the Kyoto Protocol were working to comply with their commitments, without waiting for the holdouts. Commitments varied widely. Obligations for many countries (the lesser developed) were extremely limited, such as attempting to create inventories of GHG emissions and removals. This set the stage for continued disagreements among some of the developed countries, significantly the United States, and those with increasing contributions to climate change that had no major obligations, such as China and India.

Since the Kyoto Protocol went into effect, a number of meetings and conferences of the parties have been held, as table 5.1 sets out. Meetings bring together thousands of people from governments, both elected officials and technical specialists, along with NGOs and hundreds of other interested parties. Among the most noteworthy of the outcomes are the following.

The Bali Action Plan was adopted at the thirteenth conference of the parties and the third meeting of the parties in December 2007 in Bali, focusing on cooperative action to 2012 and beyond. The Bali Action Plan aimed to have parties adopt a binding agreement in two years. During the negotiations, the EU pushed for 25–40 percent reductions in GHG emissions compared to 1990 levels. This target aimed to keep average global temperatures from rising more than 2 degrees Celsius by 2050. This target did not make it into the Bali Action Plan (Deutsche Welle 2007, as cited in Schreurs and Tiberghien 2010, 25). In hindsight, it appears the Bali Action Plan "may have been overly optimistic, and underestimated the complexity both of climate change as a problem and of crafting a global response to it" (UNFCCC 2012).

At the fifteenth conference of the parties in Copenhagen in 2009, some parties adopted the Copenhagen Accord. The accord committed $30 billion from developed countries for "fast-start financing" of adaptation and mitigation projects in developing countries, with a priority for projects in least-developed countries.

In 2010, the Cancun Agreements furthered reporting and review of GHG reductions and help to developing countries. At the seventeenth conference of the parties in Durban, the parties agreed on a second commitment period on the Kyoto Protocol and a pathway and deadlines to drawing up and committing to a new, post-2020 mitigation framework under the UNFCCC. All industrialized countries and forty-eight developing countries also affirmed their pledges up to 2020.

At the eighteenth conference of the parties in Doha, Qatar, in 2012, parties concluded their work under the Bali Action Plan. However, this meeting elicited widespread criticism for postponing action. Optimists nonetheless pointed to some forward steps. For the first time, the phrase "loss and damage from climate change" was included in an international legal document. Vulnerable nations were assured that $100 billion in public and private funds would be made available to them annually to reduce or repair loss or damage due to the effects of global warming (UNFCCC 2013b, 13 and 21-25). Remaining unanswered questions at the time of Doha were how compensatory funds would be dispersed, either prior to or after 2020, whether the United States would make funds available prior to 2020, and whether compensatory funds would be drawn from existing disaster relief or humanitarian aid budgets (EurActiv 2013).

The UN-Based Response to Climate Change: Assessment

Some nations have concluded that the UN Kyoto process cannot be effective, and a few have refused to commit to further reductions after the first commitment period or have withdrawn from the protocol. Canada did the latter in 2011. Increasingly, observers ask whether frequent international meetings on climate change are worthwhile when many countries remain in deep disagreement about committing to climate change mitigation and adaptation responsibilities.

Others see Kyoto as the only game in town. They say it is a process that helps to mobilize international cooperation and information sharing and raises the need to address questions of global climate justice, while actions to reduce the pace and magnitude of climate change and increase resilience are taken through other initiatives. Other initiatives include unilateral efforts by individual countries, regional cooperative efforts, interregional agreements, and subnational and local government actions (Climate Group 2012a). For example, thousands of cities have adopted standards modeled after the Kyoto Protocol. It is possible that many decentralized efforts to reduce emissions and improve climate resilience may not have occurred without the great visibility of climate change impacts associated with the UNFCCC and the Kyoto process.

There is also a growing sentiment that no matter what the advisability of the fundamental "architecture" of Kyoto, its processes can be streamlined. For example, mammoth meetings could be held less frequently, and the agenda for each meeting could be limited to a more narrow scope. The UNFCCC secretariat could be given additional responsibilities for Kyoto program implementation such as, say, a greater role in facilitating flexibility mechanisms or operationalizing political commitments reached outside of a more formal treaty process. Also, countries could focus on furthering the commitments of the UNFCCC through regional climate mitigation and adaptation agreements, where consensus can be developed more easily.

Other International Efforts for Mitigation

In addition to multilateral efforts through the United Nations, several countries are pursuing other programs to address climate change. The United States is involved in a number of international efforts to reduce GHG emissions. For example, the US Environmental Protection Agency (EPA) lists more than ten international and bilateral actions (US EPA 2012b), including the International Convention on the Prevention of Pollution from Ships and international cooperation on measuring the energy efficiency of computer data centers.

The United States, China, India, and seventy other parties to Annex VI of the International Convention on the Prevention of Pollution from Ships have agreed to phase in an energy efficiency design index from 2013 to 2025. By 2025, CO_2 emissions from new ships complying with the convention will be 30 percent lower than emissions from ships today. In 2011, shipping represented about 3 percent of global greenhouse emissions and was expected to grow by 150 percent to 250 percent by 2050 unless GHG-reducing policies were put into effect (US EPA 2011).

In 2011, government agencies from the United States, the European Union, and Japan adopted a standard to measure computer data center energy efficiency. The Green Grid and others continue to work to improve energy efficiency in data centers. It is estimated that data centers consumed about 2 percent of all electricity in the United States in 2010 (Glanz 2012). Data centers for Google, Adobe, eBay, and Apple are exploring fuel cells and other options to reduce the environmental impact of their energy consumption and increase the resilience of power supplies for data centers (Lesser 2012; Wesoff 2012; St. Arnaud 2012).

In 2008, more than 150 scientists from more than twenty countries signed the Monaco Declaration encouraging political leaders to promote research, study social and economic impacts, and use key findings as the basis for policies to reduce GHG emissions and reduce ocean acidification (SCOR 2008). In 2012, experts reported recent findings on responses of marine organisms to ocean acidification, impacts of ocean acidification on fisheries, and possible policy solutions (SCOR 2012).

In 2012, after the US presidential election, there was a flurry of discussion in the United States on the use of a carbon tax to reduce GHG emissions (Borenstein 2012). A renewed interest in looking for ways to avoid more frequent occurrences of super storms helped motivate the discussion. A super storm, Sandy, struck the eastern United States just before the election, causing extensive damage in New York, New Jersey and neighboring states. British Columbia has had a revenue-neutral carbon tax in place since 2008. As of July 2012, the tax is $30 per metric ton of CO_2-equivalent emissions (British Columbia Ministry of Finance, 2012, 66–68). Norway and Australia have carbon taxes (Carbon Tax Center, 2012), and a number of other countries and subnational jurisdictions have carbon taxes as well.

European Union

The European Union, representing almost 30 percent of 1990 emissions from industrialized countries, led in implementing policies to reduce GHG emissions before the Kyoto Protocol went into effect. In a 2003 survey conducted by the European Community, 88 percent of European voters supported taking immediate actions to address climate change (European Commission 2003).

By signing the Kyoto Protocol, countries in the European Union agreed to reduce their GHG emissions by 8 percent of 1990 levels from 2008 to 2012, although targets for individual EU countries vary. The EU Parliament has made this goal legally binding, and a number of regional and national policies—including the creation of a trading system for CO_2 emissions—aim to reduce GHG emissions and help achieve this target. In 2012, the European Commission reported that the fifteen European countries with Kyoto Protocol obligations were on track to meeting their commitments (European Commission 2012, 3).

Without waiting for others, the European Union is implementing binding legislation to reduce GHG emissions to 20 percent below 1990 levels. Legislation to achieve this goal includes measures to, by 2020, improve energy efficiency by 20 percent and use renewable energy to produce 20 percent of energy consumed; binding targets to reduce CO_2 in the transportation sector; the EU Emissions Trading System; and other measures (European Commission 2013b).

At the same time, the European Union continues to encourage other major emitters around the world to "undertake their fair share of a global emissions reduction effort." As part of its negotiations with other countries, the European Union has offered to reduce GHG emissions to 30 percent below 1990 levels by 2020. In addition, the European Union has published a roadmap for reducing its GHG emissions 80–95 percent below 1990 levels by 2050 (European Commission 2013b).

Buoyed by strong public support for action on climate change, governing bodies of Europe, including the European Commission, European Parliament, and the rotating EU Presidency, have reinforced one another's leadership on this issue. For example, there has been pressure on the country rotating into the EU presidency to maintain or improve on the climate change policies of the prior presiding country (Schreurs and Tiberghien 2010, 26).

The United Kingdom has taken a leading role in addressing climate change. The position of the Blair government was that climate change is an urgent matter, action is needed now to avoid disaster, and policies must encourage investment in science and technology and behavior changes to reduce GHG emissions and expand the economy. The United Kingdom was the first country to set a target that was more aggressive than the Kyoto Protocol—reducing carbon dioxide emissions by 20 percent by 2010. It has also set a goal of reducing carbon dioxide 60 percent by 2050. To help achieve these goals, a controversial climate-change levy was adopted.

In addition to cutting emissions produced in the United Kingdom, concern is rising about the need to cut emissions embodied in goods consumed in the United Kingdom but produced elsewhere. In April 2013, a study by the UK Committee on Climate Change reported that taking GHG emissions used to produce imported goods into account, UK GHG consumption emissions have increased by about 10 percent since 1993. To reverse this trend, the committee called for a global agreement to greatly reduce emissions: "An ambitious and comprehensive global deal driving new policies will be essential so that global emissions are reduced in a manner consistent with the climate objective (e.g. such that deep cuts in global emissions are achieved by 2030), as a consequence of which UK imported emissions would fall" (UK Committee on Climate Change 2013, 9).

Germany and the Netherlands are also taking innovative steps to reduce GHG emissions, including greatly increasing their use of renewable energy, particularly offshore wind energy. The Dutch are even controlling the greenhouse gases from their massive flower-producing greenhouses: they have a requirement that all new greenhouses must be climate-neutral by 2020, and there are many examples of innovative and integrated systems that achieve this goal (Government of the Netherlands 2013a; Government of the Netherlands 2013b). Germany has a Kyoto agreement target of 21 percent below 1990 levels by 2012 and has set an additional target of 40 percent reduction of CO_2 by 2020. In 2000, Germany planned to close down its nineteen nuclear power plants by 2020. Since 2000, German plans for nuclear phase-out have gone through stops and starts. After the disaster at the nuclear power plant in Fukushima, Japan, Germany renewed its commitment to shut down its nuclear power plants. The current phase-out date is 2022. (Davidson 2012; Giddens 2011, 77-78). Germany is aggressively developing renewable energy. In 2010, Germany adopted a goal aiming for 60 percent of gross final energy consumption from renewable resources by 2050, including about 80 percent renewable electricity by 2050 (Federal Republic of Germany 2010, 5). In February 2013, the German federal environment ministry reported GHG emissions had dropped more than 25 percent from 1990 levels. However, the ministry noted that emissions increased 1.6 percent in 2012 compared to 2011 due to increased use of coal to generate electricity. The head of the ministry called for a reduction in the surplus of carbon allowances in the EU emissions trading system. This would help raise the cost of electricity from coal compared to lower-carbon sources (Nicola 2013). In April

2013, the European Parliament voted against reducing the number of surplus allowances (Grose 2013).

European Trading System for GHG Emissions

Those who look to market mechanisms as a strategy for addressing climate change often look to carbon taxation or cap-and-trade systems to place a price on carbon emissions. The European Union has been a leader in the latter, although surplus allowances in its cap-and-trade system have caused prices to drop.

The European Commission describes its EU emissions trading system (ETS) as "a cornerstone of the European Union's policy to combat climate change and its key tool for reducing industrial greenhouse gas emissions cost-effectively" (European Commission 2013a, 1). Europe's ETS was the first, and is the largest, international system for trading GHG emission allowances. It covers more than 11,000 power stations and industrial plants in thirty-one countries (including EU, Iceland, Liechtenstein, and Norway).

As with other cap-and-trade approaches, the European system sets a limit on the total amount of certain GHGs that can be emitted by the factories, power plants, and other installations in the system. The cap is reduced over time so that total emissions fall. In 2020, emissions from sectors covered by the EU ETS will be 21 percent lower than in 2005. Within the cap, companies receive or buy emission allowances, which they can trade with one another as needed. They can also buy limited amounts of international credits from emissions-saving projects around the world.

The limit is intended to ensure that credits actually have value. "After each year a company must surrender enough allowances to cover all its emissions, otherwise heavy fines are imposed. If a company reduces its emissions, it can keep the spare allowances to cover its future needs or else sell them to another company that is short of allowances" (European Commission 2013a, 1). The idea is that emissions are cut where it costs least to do so. The EU argues that "in allowing companies to buy international credits, the EU ETS also acts as a major driver of investment in clean technologies and low-carbon solutions, particularly in developing countries" (European Commission 2013a, 1–2).

The program is now in its third phase, running from 2013 to 2020. A 2009 revision made important changes. It set a single EU-wide cap on emissions and it made auctioning the default way to allocate allowances not given away for free. Also, the new phase added more sectors and gases (European Commission 2013a).

The poor economy throughout Europe over the past few years, and a declining perception of the importance and urgency of climate change in some sectors, have pressured Europe to reconsider its commitments. The European Commission postponed (or "back-loaded") the auctioning of some allowances as a short-term response and launched a discussion on measures to provide a structural solution to reform in the market in the longer run. In 2013 carbon prices dropped to around 3 euros; some analysts believe a price of 50 euros is necessary for the market to be effective (Reuters 2013).

Japan

In the Kyoto Protocol, Japan agreed to cut GHG emissions to 6 percent below 1990 levels by 2012. The fact that a Japanese city hosted the negotiations and lent its name to the resulting agreement was an important consideration in Japan's political dynamics leading to ratification of the Kyoto Protocol. The Kyoto Protocol represents an important symbol of Japan's efforts to become a leader on climate-change issues, and continues to spur action on climate change in Japan (Tiberghien and Schreurs 2010, 140–141).

In 2002, the government of Japan reported that about 90 percent of its GHG emissions came from the combustion of fossil fuels for energy. Japan also reported the highest energy efficiency in the world, so reaching its Kyoto targets will not be easy.

One Japanese policy is to develop new products to help business, individuals, and the government to conserve energy. Voluntary action by industry is central. To encourage residential conservation, the government is promoting technology that displays the cost of energy as it is being used in the home (Government of Japan 2002, 72–78).

In December 2010, Japan decided to put plans to start carbon trading on hold. The plans called for a cap-and-trade system to be launched in Japan in 2013. Concerns were raised that the cap-and-trade program in Japan could place industry at a disadvantage compared to competitors in India and China, countries that did not face similar restrictions. The primary reason the Japanese government postponed introduction of cap and trade was because the government decided to introduce a carbon tax, and industries strongly objected to introducing two economic instruments that work toward the same objectives. The first carbon tax in Japan was introduced in 2012 (Kameyama 2013).

The Japanese government advocates switching from fossil fuels to renewable energy. In the past, it has also called for expanding nuclear energy as a part of its plans for reaching its GHG emission targets under the Kyoto Protocol. But as of 2002, public opinion in Japan was opposed to expansion of nuclear energy (Government of Japan 2002, 98).

In 2011, an earthquake and tsunami led to release of nuclear radiation from a nuclear power plant in Fukushima, Japan (Tsukimori 2012). Prior to the disaster, nuclear energy provided about 30 percent of Japan's electricity (Tabuchi 2012). As part of its response, Japan has put policies in place for rapid expansion of small and medium-scale renewable energy facilities. As of October 2012, Japan is ahead of schedule to achieve its goals for renewable energy. The driver for the speedy response is the attractive price that utilities offer for electricity from rooftop solar and other distributed renewable energy facilities (Burger 2012).

Regarding GHG emission reductions, influential energy-intensive industries in Japan object to targets beyond those affecting important economic competitors in other countries (Kameyama 2012, Hattori 2007). This concern regarding competitiveness affects Japan's position at the international level. Japan decided and announced at COP17 in 2011 that it would not participate in the second commitment period of the Kyoto Protocol, although it would remain a party to the protocol. Japan has been unhappy that the Kyoto Protocol did not cover emissions from major emitters, mainly the United States and China (Kameyama 2013).

China

China ranks among the world's top five GHG emitters by some measures and first in emissions of leading contributing sources of climate change. Per capita emissions, historically much less than in Annex I countries, have reached equivalence with averages in the European Union, although there is uncertainty in these figures (PBL Netherlands Environmental Assessment Agency 2012). China has signed the Kyoto Protocol and participates in the aspects of the agreement that apply to developing countries, including updates of national inventories of human-caused GHG emissions, programs to help lessen the amount or speed of climate change, and programs to adapt to climate change. China is one of the most active developing countries in Clean Development Mechanism projects (United Nations Environment Programme 2013). China is becoming more involved in the IPCC and the Clean Development Mechanism Executive Board (Heggelund et al. 2010, 250). However, the Kyoto Protocol has no mandatory GHG reduction targets for developing countries.

According to the International Energy Agency (IEA), CO_2 emissions from coal in China were less than 2,500 million metric tons in 2000 and 2001. In 2010, this increased to more than 6,000 million metric tons, which was more than 45 percent of the world's total CO_2 emissions from coal for 2010 (IEA 2012). In 2011, China was a net importer of coal; imports were about 5 percent of China's total coal consumption (IEA 2013a).

China is taking significant actions to reduce the rate of growth of energy and carbon per unit of GDP. However, a 2011 study by the Lawrence Berkeley National Laboratory points to a continued increase in total CO_2 emissions from China unless China's energy and carbon intensity goals through 2020 are met and additional carbon-reducing measures are put in place (Zhou et al. 2011).

In its eleventh five-year plan, China set an energy-intensity reduction goal to reduce energy consumption per unit of gross domestic product by about 20 percent from 2006 to 2010. China reports achieving a reduction of about 19 percent in

2010 compared to 2005 (National Bureau of Statistics of China 2011).

For 2011 through 2015, China has a goal to reduce energy intensity by 16 percent and reduce carbon intensity by 17 percent. For 2020, China's goal, outlined in the Copenhagen Accord, is to reduce carbon intensity below 2005 levels by 40 percent to 45 percent (Zhou et al. 2011).

However, it is unlikely that China will substantially reduce its reliance on coal as a source of energy for some time to come. In an interview conducted in February 2013, Zou Li, Deputy Director of China's National Center for Climate Change Strategy, indicated that China is aware of the GHG benefits of moving away from coal for the production of energy, but plans to continue relying on coal to promote economic development: "If China could replace coal with oil as a primary source of energy, emissions would drop by one third. If we could replace coal with natural gas, they would drop by two thirds. But China's main resource is coal. We only have limited amounts of other sources of energy, and obviously a reliance on imports is unrealistic" (Xu and Zhang 2013, 3). In addition, China is a developing nation, and although development has progressed rapidly in major cities, "the welfare of millions of rural residents isn't yet assured. Healthcare, unemployment benefits, pensions, all of these are weak. Many Chinese people have no safe drinking water, and our per-capita GDP ranks ninety-something globally" (Xu and Zhang 2013, 2).

Although China plans to continue using coal to fuel economic development, in September 2013 additional steps were announced to put limits on burning coal, along with removing high-polluting vehicles from the road (Wong 2013). And there are indications that China is striving to increase development of renewable electricity resources and renewable fuels as well.

In 2007, China adopted a plan for renewable energy with a target for renewable electricity and renewable fuels to provide 10 percent of energy consumption by 2010 and 15 percent by 2020. The plan includes individual targets for hydropower, wind, solar photovoltaic, solar water heating, and a number of bioenergy resources. China surpassed the 2010 targets for all but biogas consumption and nongrain fuel ethanol. Expansion of wind and solar photovoltaic installations exceeded the 2010 targets many times over (Chang et al. 2012, 2–3). In April 2013, China led the world in renewable energy investments (Pew Charitable Trusts 2013).

China is highly vulnerable to climate change and has placed growing emphasis on preparing for potential impacts (Heggelund et al. 2010, 236). In 2007, China also published a plan for climate change, including adaptation. Key activities for climate adaptation included improvements in water management infrastructure and irrigation technologies, development of hardy strains of crops, strengthened coastal infrastructure, expansion of coastal mangrove forests, construction of an early-warning system for tidal disasters, and improved forest protection policies (National Development and Reform Commission, China 2007). In 2011, China announced a target to increase total forests in China by more than 21 percent by 2015 (Lewis 2011).

China is in the process of taking steps to improve resilience to climate change. For example, in 2012, a five-year climate adaptation project was completed in the key agricultural area of China. The World Bank and the Global Environment Facility (GEF) funded the project. The project aimed to demonstrate irrigation and water-management measures in the Huang-Huai-Hai Basin (the 3H Basin), which is home to 400 million people (GEF 2012a; GEF 2012b). The 3H Basin provides half of the grain produced in China. However, water resources are in short supply and conditions are expected to worsen with climate change. The GEF expects the irrigation and water-management measures demonstrated in this project will be replicated widely in the basin and other rural areas of China (GEF 2012b).

China's coastal ecosystems also continue to face challenges. In 2012, results from a survey by China's State Oceanic Administration indicated severe deterioration in China's coastal ecosystems, especially coastal mangrove forests, coral reefs, and wetlands. The survey also indicated that fisheries are suffering from rising levels of ocean acidity and pollution, including a brown tide lethal to shellfish (Qiu 2012).

The United States: The Federal Government

The United States produced over 36 percent of the 1990 emissions in the industrialized countries (the Annex 1 countries), and its position on climate-change law and policy has shifted and vacillated over the last three decades. Table 5.2 lists some of these positions.

In March 2001, the United States reneged on its earlier support of the 1997 Kyoto Protocol. Shocking to some, this was no surprise to close observers of environmental policy. As early as July 25, 1997, the US Senate had passed the Byrd-Hagel resolution (Senate Resolution 98) by 95 to 0 to pressure Kyoto negotiators for a global agreement rather than one that bound only industrialized countries. The resolution indicated that Senate support would be withheld if (1) the agreement reached at Kyoto imposed binding limits on industrialized countries without also imposing binding limits on developing countries or (2) the agreement would result in serious harm to the economy of the United States.

The United States signed the Kyoto Protocol in November 1998, but President Bill Clinton did not send it to the Senate for ratification. Although the agreement advanced international efforts to address climate change, it failed to comply with the Byrd-Hagel requirement that developing countries would have to limit greenhouse gases.

Some developing countries had recently joined the OECD, a status that is relevant for some climate-change commitments, but they had not yet taken responsibility for controlling their GHG emissions. Furthermore, some had very low, stable levels of GHG emissions, and others, such as China, had rapidly growing emissions.

To lower emissions of greenhouse gases, the United States continued to negotiate bilateral agreements with developing countries, including China. Its international efforts, such as contributions to an international institution for environmental improvements called the Global Environment Facility, continued, but Clinton considered it premature to submit the Kyoto Protocol to the Senate for ratification because it did not require binding commitments from developing countries.

In the 1990s, the US government faced strong pressure from industry to oppose the Kyoto Protocol. A group named the Global Climate Coalition attended negotiations, contributed to the IPCC scientific assessment documents, provided comments on proposed government programs on climate change, and lobbied Congress. According to this coalition—which included large and small businesses in agriculture, forestry, electric utilities, railroads, transportation, manufacturing, mining, oil, and coal—the Kyoto Protocol was unfair because developing countries were not required to reduce emissions, US growth could be severely hampered if energy prices for consumers increased dramatically as a result of its requirements, and its

Table 5.2
The US Response to Climate Change

Date	Federal Action
1980s	Gore-Wirth legislative initiatives call for increased funding on climate research and energy conservation and renewable energy.
1987	Congress passes the Global Climate Protection Act, which calls for climate-change policy to be coordinated nationally.
1992	George H. W. Bush attends the conference where the United Nation Framework Convention on Climate Change is adopted in Rio de Janeiro.
1992	The United States signs and ratifies the Framework Convention on Climate Change.
1992	Congress passes the Energy Policy Act; it sets rules for restructuring electricity delivery, reducing American dependence on foreign oil, and requiring electricity-generating utilities to report annually on carbon dioxide emissions.
1995	The Clinton administration endorses the Berlin Mandates calling for development of binding emission reductions.
1997	The Byrd-Hale resolution passes 95 to 0 and recommends that no future climate-change agreements be signed without commitments from developing countries.
November 1998	The Kyoto Protocol is signed but is not submitted to Congress.
1998–2013	Bilateral agreements with developing countries on climate-change emissions; participation in Global Environmental Facility projects.
March 2001	Bush administration disengages from the Kyoto Protocol.
2002	The Bush administration focuses on carbon-intensity reductions and Voluntary Innovative Sector Initiatives Opportunities Now (VISION).

Table 5.2 (cont.)

Date	Federal Action
2003	The McCain-Lieberman bill proposes that GHG emissions be reduced to 2000 levels by 2010 and is defeated 56 to 44.
2005	Energy Policy Act passes, encourages carbon capture, sequestration and storage.
	Secretary of Energy mandated to implement 10-year carbon capture research and development program for carbon dioxide capture technologies on combustion-based systems.
2007	Energy Independence and Security Act passes, which also encourages carbon capture, sequestration and storage.
	Supreme Court finds, in *Massachusetts v. EPA*, that the EPA had abused its discretion in its failure to regulate GHG emissions under the Clean Air Act. Finding creates basis for new climate change regulation.
	President Barack Obama calls for effective climate change regulations in his 2007–2008 presidential campaign.
2008	President Obama directs Congress to pass a comprehensive climate bill in its 2009 term. Congress considers several pieces of legislation addressing GHG emissions but fails to pass more than a few.
	Consolidated Appropriations Act passes, mandating a "comprehensive and effective national program of mandatory market-based limits and incentives ... that slow, stop and reverse the growth of emissions." The act also identified funding for EPA's promulgation of a rule requiring mandatory reporting of GHG emission.
2009	Omnibus Appropriations Act passes, mandating EPA to develop a GHG reporting rule: approximately 85 percent of GHG emitters in the United States are thus required to report their emissions.
	American Recovery and Reinvestment Act passes, provides substantial funds for investment in clean energy technology.

Date	Federal Action
2009 (cont.)	President Obama creates the Interagency Climate Change Adaptation Task Force through Executive Order.
	President Obama creates the Interagency Ocean Policy Task Force (IOPTF) through Executive Order; IOPTF releases Interim Report for public comment.
	President Obama issues "GreenGov Challenge," engaging federal employees in greening the government.
	American Clean Energy and Security Act passed by House but does not pass the Senate. The act would have required establishing a national cap-and-trade system for GHG emissions. Would also have required a reduction in GHG emission levels to 83 percent of 2005 levels by 2020.
2010	Worsening economic climate and changes in political picture result in diminished support for climate-change legislation.
	In *American Electric Power v. Connecticut*, the Supreme Court finds that GHG litigation should take place under the Clean Air Act and not under federal common law/public nuisance.
	EPA denies all ten petitions challenging its Endangerment and Cause-or-Contribute Findings on the basis of climate science validity.
	EPA issues "tailoring rule" establishing permitting requirements for GHG emissions and begins to regulate stationary sources. Issues final rule estimating threshold GHG permit requirements for new and existing GHG-producing facilities under New Source Review, Prevention of Significant Deterioration, and Title V.
	Senate rejects resolution to eliminate EPA's ability to regulate GHG emissions.
	Proposed cap-and-trade legislation fails to pass Congress.
	EPA and the National Highway Traffic Safety Administration jointly issue first national rule limiting GHG emission from light-duty trucks and cars.

Table 5.2 (cont.)

Date	Federal Action
2010 (cont.)	Poor economic recovery results in several attempts to delay implementation of GHG emissions regulations by representatives from coal-producing states.
	President Obama establishes the Interagency Task Force on Carbon Capture and Storage (CCS) through executive order; the task force delivers recommendations to the president for comprehensive, cost-effective deployment of CCS by 2020.
	Council on Environmental Quality issues draft guidance for federal GHG reporting and accounting.
	Ocean Policy Task Force issues final recommendations. President Obama establishes National Ocean Policy and National Ocean Council through executive order.
	President Obama expands GHG reduction target for federal operations.
2011	EPA begins to finalize its approach to regulating nonstationary sources of GHG emissions under the CAA.
	Obama administration announces grid modernization initiative to foster a clean energy economy and spur innovation and unveils grid modernization pilot projects also aimed at creating jobs.
2012	Interagency Ocean Policy Task Force releases its National Ocean Policy Implementation Plan.
	US Global Change Research Program (USGCRP) releases its National Global Change Research Plan 2012–2021.
	USGCRP releases Climate Change Indicators in the United States, 2012 report.
	Obama administration releases action plan to address ocean challenges.
	Obama administration and Great Lakes states announce agreement to foster development of offshore wind projects.
2013	National Institutes of Health investigates impacts of climate change on human health; identifies nearly 800 projects focusing on climate change and/or its effects.

Sources: US Global Change Research Program 2012, 2013; White House Council on Environmental Quality 2013a, 2013b, and 2013c.

targets and timetables would gravely harm American families, workers, older people, and children (Global Climate Coalition n.d.; SourceWatch n.d.). The coalition disbanded once it believed that the Bush administration was set to change the US position on the Kyoto Protocol.

In March 2001, President George W. Bush rejected the Kyoto Protocol, saying "it exempts 80 percent of the world, including major population centers such as China and India, from compliance, and would cause serious harm to the US economy" (White House 2001, 1). Bush said that the United States generates more than half of its electricity from coal and that caps on CO_2 emissions would shift electricity generation from coal to natural gas, which would raise energy costs: "At a time when California has already experienced energy shortages, and other Western states are worried about price and availability of energy ... we must be very careful not to take actions that could harm consumers. This is especially true given the incomplete state of scientific knowledge of the causes of, and solutions to, global climate change and the lack of commercially available technologies for removing and storing carbon dioxide" (White House 2001, 1).

Californians and Americans in general do not agree with this statement. In April 2001, ABC reported on a survey indicating that 61 percent of Americans, including 52 percent of Republicans, rejected the arguments made by Bush and said the United States should ratify the Kyoto Protocol (Sussman 2001; ABC-News.com 2001). In a 2002 Harris Poll, more than 50 percent of those who had heard of climate change disagreed with Bush's rejection of the Kyoto Protocol (Harris Interactive 2002).

In March 2013, a Gallup Poll found that more than 60 percent of Americans believe scientists agree global warming is occurring. The poll also found that more than 60 percent believe global warming will not pose a serious threat to them personally during their lifetime. Nonetheless, more than half of Americans report worrying about global warming a great deal or a fair amount (Saad 2013). In January 2013, a survey of Republicans and Republican-leaning independents indicated that over three-fourths support greatly expanding the use of renewable energy and over two-thirds of those holding this view support expanding the use of renewable energy immediately. The survey also found that a majority of respondents think climate change is occurring and support US action to address climate change (Maibach et al. 2013).

A bipartisan Senate bill introduced in 2003 proposed that GHG emissions be reduced to 2000 levels by 2010. The McCain-Lieberman bill required that a cap be reached through on-site measures or by the trading of GHG emission rights. This bill failed by a vote of 55 to 43. Opponents argued that carbon dioxide poses no direct threat to public health and that the McCain-Lieberman requirements would burden families and communities. Supporters said that climate change is real and that problems such as the loss of sea ice in the Arctic would worsen until society reduces GHG emissions. Later Lieberman bills with co-authors, including those seeking support from the business community by adding hundreds of millions of dollars in subsidies for cleaner electrical energy and nuclear energy, also failed to move forward.

Rather than reduce total GHG emissions, in 2002 the Bush administration adopted the Global Climate Change Initiative, which aims to reduce the ratio of GHG emissions to economic output by 18 percent by 2012 through domestic voluntary actions and continued research on climate change. Reaching the target was compared to reducing emissions to 5 percent below 1990 levels by the 2008–2012 commitment period of the Kyoto Protocol. According to the Bush administration, the initiative "ensures that America's workers and citizens of the developing world are not unfairly penalized."

The US position emphasizes research. The federal government provides about $3 billion per year for continuing research on climate change. Support for this amount was written into law in 1990, and in the president's budgets for later years, including fiscal years 2004 and 2005, federal agencies reported plans for continued research. The 2014 budget would provide $2.7 billion for the thirteen-agency Global Change Research Budget (USGCRP), 6 percent more than the amount enacted for 2012. The focuses are on prediction, mitigation, and adaptation to global change, including climate change (Holdren 2013, 10–11). For fiscal year 2015, one of the priorities within adaptation work is to "enhance utility of data and tools for purposes such as catastrophe risk management in a non-stationary climate" (Burwell and Holdren 2013, 4). These figures have been fairly stable in recent years with a slight spike in 2009 based on the Recovery Act contribution. The US program includes bilateral and regional research with developing and developed countries.

Congress has taken some initiatives addressing climate change. In June 2005, the Senate passed a resolution stating its intention to require, at some future date, a program of mandatory GHG limits and incentives: "It is the sense of the Senate that Congress should enact a comprehensive and effective national program of mandatory, market-based limits and incentives on emissions of greenhouse gases that slow, stop, and reverse the growth of such emissions" (US Senate 2005, 1271). As of 2013, the economic slowdown and continued resistance by some Congressional leaders has resulted in only a few initiatives, such as passage of the Energy Independence and Security Act—as shown in table 5.2.

The Federal Government and GHG Regulatory Actions

In 2007, the US Supreme Court decided in *Massachusetts v. EPA* that the EPA had abused its discretion in failing to regulate GHG emissions under the Clean Air Act, thereby paving the way for the creation of new federal climate change regulations. Since then, the federal government has not supplemented the very limited legislation that directly addresses climate change. Instead federal regulatory actions aimed at GHG emissions are occurring under the Clean Air Act, often influenced by threatened or actual litigation (Osofsky and McAllister 2012).

Federal legislation related to climate change has focused mainly on the energy sector (Osofsky and McAllister 2012). For example, in 2007, Congress passed the Energy Independence and Security Act (EISA), which authorizes research into and testing of technologies for carbon capture, storage, and sequestration.

In 2008, newly elected President Barack Obama sought a comprehensive climate bill from Congress during its 2009 term. The bills related to climate change enacted into law during this timeframe include the Consolidated Appropriations Act (2008), the Omnibus Appropriations Act (2009), and the American Recovery and Reinvestment Act (2009). The Consolidated Appropriations Act includes a call for a "comprehensive and effective national program of mandatory, market-based limits and incentives on emissions of greenhouse gases that slow, stop, and reverse the growth of such emissions." The 2009 Omnibus Appropriations Act mandated the EPA to develop a GHG reporting rule for certain stationary facilities. The American Recovery and Reinvestment Act, known better as the Stimulus or the Recovery Act, provided incentives for investment in clean energy technologies (Osofsky and McAllister 2012).

In 2010, as Congress turned its attention to the nation's economy, its focus on climate change legislation faded. Support for further GHG regulation diminished (Broder 2010). Although numerous bills, including cap-and-trade bills, were proposed in Congress, most proposals never made it out of committee.

In 2011, the Supreme Court heard and decided another case related to climate change, *American Electric Power v. Connecticut*. The court's decision in that case reinforced the Clean Air Act regulatory approach to GHG emission reduction and simultaneously prevented regulatory attempts based on federal common law nuisance suits. However, the court also found that, should Congress abolish EPA's authority to regulate GHG emissions under the Clean Air Act, federal common law nuisance suits would be allowed.

The EPA's Regulation of GHG Emissions

In 2009, the EPA began to finalize its approach to regulating GHGs under the Clean Air Act. It first found that current and projected atmospheric concentrations of six major GHGs may reasonably be believed to endanger the public health and welfare of present and future generations (EPA 2009).

The agency also found that the emission of these gases from new light-duty motor vehicles is endangering public health and welfare. This finding was used as a basis for regulating motor vehicle GHG emissions. Subsequent to the EPA's denial of all ten petitions challenging its Endangerment and Cause-or-Contribute Findings (findings under the statutory framework related to threats to public health and welfare and effects on atmospheric pollution) on the basis of climate science validity in 2010, numerous other petitions for review or reconsideration of various EPA actions have been filed, and legal battles continue in the courts.

In 2010, EPA and the National Highway Traffic Safety Administration (NHTSA) jointly issued the first national rule limiting GHG emissions from 2012 to 2016 model year light-duty trucks and cars and began working on regulations for vehicles to be produced between 2017 and 2025. The collaborative EPA-NHTSA rule-making efforts, known as the "National Program," represent the first time that fuel efficiency standards and vehicle tailpipe emissions standards have been regulated on a combined, rather than separate, basis. In 2011, the EPA and the NHTSA adopted the Heavy-Duty National Program, which regulates GHG emissions from medium-and heavy-duty vehicles. In the same year, the Obama administration granted a waiver to the state of California to allow stricter standards for vehicle emissions in the state. Other states may choose to follow federal standards or the stricter California standards (Osofsky and McAllister 2012).

In 2010, the EPA issued a "tailoring rule" that establishes permitting requirements for GHG emissions by detailing statutory thresholds under sections of the Clean Air Act related to prevention of significant deterioration (PSD) of air quality that meets national standards and other sections of the act. Also in 2010, the EPA expanded its regulatory efforts to stationary sources of GHG emissions. The agency established threshold GHG-permitting requirements for new and existing GHG-producing facilities. In 2011 and 2012, the agency conducted a rulemaking process directed at the development of permitting requirements for stationary sources (Osofsky and McAllister 2012).

Interest Group Politics

Interest groups affect nations' reactions to climate change and their decisions on entering international agreements. Some OPEC nations have taken very strong positions limiting international regulation of fossil fuel use, whereas Norway, a large oil exporter, is highly committed to reducing greenhouse gases at home (Dolsak 2001).

In the United States, the national position on climate change develops through successive sessions of Congress and presidential administrations, with input from environmental groups, industry groups (including oil, coal, nuclear, and alternative energy), transportation companies, farmers, ideology-based public-interest organizations, labor unions, cities, counties, and states.

Some lobbyists focus specifically on climate change, while others promote general positions on government regulations, some of which influence climate. Organizations may materialize for a particular battle; others are more or less permanent throughout environmental controversies. Groups may come together for strategic purposes, promoting an action that seeks to decrease emissions while at the same time creating jobs, for example. The outcome is an "anguished, often moralistic, rhetoric that has polarized national debate and made any semblance of consensus at that level ... elusive" (Rabe 2002, 23).

Partisan politics can explain some of, but not the entire, story of the official US position on climate change. Democrats generally have supported official cooperation in international agreements on climate change, and Republicans generally have been skeptical if not downright opposed, but this breakdown does not tell the whole political story. Although Al Gore promoted international responses to the problem (for example, by attending the Kyoto Protocol meeting and publishing his book *Earth in the Balance*), the Clinton administration backed away from a Gore proposal to set a tax on the use of fossil fuels, and when Congress turned Republican in November 1994, "climate change policy moved even further to the recesses" (Rabe 2002, 12).

Still, a few Republicans have taken vocal positions promoting federal action on climate change. Some interest groups that are critical of government regulations have recognized that tight controls on climate gases might result in profits for certain industries. Other political groups, such as conservative Christians, see climate change as fundamentally an ethical or moral issue. Those that emphasize international affairs hope for liberation from Middle East dominance of the energy industry if oil becomes a less important part of the energy sector. With a split Congress under the administrations of President Obama, interest groups have had some success in promoting activities that lead to outcomes aimed at climate change mitigation and promoting adaptation. Other groups remain successful in ensuring that the term *climate change* is rarely articulated in political campaigns and in the halls of Congress. When the president utters the term it is considered a major action, one of political courage. Groups like the Heartland Institute repeat that, "evidence continues to amount that fears of man-made global warming are over-blown and not supported by the latest science" (Goreham 2013). For further discussion of this topic, see chapter 4 by Naomi Oreskes in this volume.

As Bryner and Duffy have summarized:

A concerted campaign by Republicans and industry officials to sow doubt about climate science contributes to the difficulty of generating support for climate policies ... [while] ideological opposition to climate change policy ... means more government regulation. Strong opposition to environmental politics from businesses that oppose new regulatory burdens in general and climate change policy in particular has also been a factor. Companies such as Exxon and Koch Industries have invested in advocacy groups that have raised questions about climate science. ... Because of the major role of fossil fuels in GHG emissions, climate policy is also energy policy, and efforts to transform energy production from fossil fuels to cleaner alternatives are imperiled by the political influence of oil, gas and coal interests. (2012, 51–52)

Some consider climate policy impeded by its consideration in an "era of base politics ... where strident ideological and partisan conflict has replaced bipartisanship and political parties routinely use environmental conflict as a mean of mobilizing their base" (Goreham 2013, 52; citing Richard N. L. Andrews).

Interest groups' positions on strategies to address climate change have shifted depending in part on the economic benefits associated with a strategy at a particular time. For example, at the time of the negotiations for the Kyoto Protocol,

The reasons why government actors backed carbon trading include a positive institutional memory regarding emissions trading (the US Administration), the prospect of a significant revenue stream from auctioning permits (the US Congress), and strategic concerns related to an international agreement on a climate treaty and EU-internal power dynamics (the European Commission). As for the NGO community, business-oriented environmental groups pushed the emissions trading agenda because they were able to mobilize business support for it as opposed to alternative mandatory climate policies. (Meckling 2011, 174)

The States and Cities

Federal policies do not tell the whole story on US climate change. Public opinion is generally supportive of efforts to address climate change, and many state governments and cities have passed laws or adopted policies addressing this issue.

In 2003, more than three-quarters of Americans thought the United States should reduce GHG emissions (Gallup 2003). In March 2012, more than three out of five Americans supported "imposing mandatory controls on carbon dioxide emissions and other greenhouse gases" (Gallup 2012).

The fifty US states have varying views on climate change. At least sixteen states (including Alabama, Kentucky, South Carolina, and Virginia) have passed resolutions asking the federal government to reject the Kyoto Protocol (Rabe 2002, 20), while others have taken actions that go far beyond its commitments. In 2003, the attorneys general of Connecticut, Maine, and Massachusetts filed a federal lawsuit alleging that the EPA under the Bush administration had failed to implement the Clean Air Act to include regulation of carbon dioxide. Under the Clean Air Act, the EPA is required to review and, where appropriate, revise regulations based on scientific information about the environmental health effects of a proposed targeted pollutant. The 2003 lawsuit asserted that carbon dioxide is a pollutant "that causes global warming with its attendant adverse health and environmental impacts" and called on the Bush administration to revise the Clean Air Act to include regulation of carbon dioxide (Rabe 2002, 164–65). In April 2007, the US Supreme Court, by a vote of 5 to 4, ruled in favor of the states.

Many states began to take action earlier. For example, a number of New England governors and eastern Canadian premiers developed a regional Climate Change Action Plan in 2001. The group committed to reducing GHG emissions to 1990 levels by 2010 and to at least 10 percent below 1990 levels by 2020 (Conference of New England Governors and Eastern Canadian Premiers 2001). In addition, they agreed to reduce emissions sufficiently over the long term to eliminate any dangerous threat to the climate, which scientists estimate to be 75 to 85 percent below current levels. Other states have passed laws that create mandatory cap-and-trade programs, set GHG reduction targets, and require GHG inventories and reports. Also, some states have laws that promote the capture and storage of carbon dioxide in trees, underground, and in other locations and that strengthen energy efficiency, conservation, and renewable energy in electricity generation and transportation (Center for Climate and Energy Solutions 2011).

By 2011, almost half of US states had statewide emission targets and goals aimed at reducing greenhouse gases (Center for Climate and Energy Solutions, 2011). As early as 1989, New Jersey had an executive order that called on its state government to lead in decreasing emissions of climate-change gases. In 1998, New Jersey, under Governor Christine Whitman (who later became the US EPA administrator), pledged to reduce emissions in a way that would reach Kyoto Protocol goals within a few years. However, a 2013 analysis indicates New Jersey's GHG emissions for 2010 are similar to emissions in 1990 (ENE 2013). Another Republican governor, George W. Bush, signed the Texas Public Utility Regulatory Act of 1999, which required state utilities to move toward greater use of renewable energy sources. While Texas continues to be a major source of GHG emission, its recent use of renewable energy has increased. In 2000, Nebraska passed a program designed to sequester carbon through agricultural practices. Several nearby states followed suit. A number of states, including New Hampshire and Massachusetts, have applied a cap on carbon dioxide and other pollutants. Oregon has linked the siting of new energy plant facilities to commitments to reduce climate-change gas emissions.

Beginning in 2010, the EPA has required facilities to publicly disclose releases of carbon dioxide on a yearly basis; this applies to all large direct-emission and fuel supply sources. The data for 2010 show that the sectors with the largest GHG emissions are power plants and petroleum refineries (EPA 2012a).

In many states, the use of coal to generate electricity is declining because of environmental concerns and the low price of natural gas (Union of Concerned Scientists 2012). Hydraulic fracturing (fracking) of shale, a common type of underground rock formation, has increased the availability of natural gas in the United States. The increased supply has caused the price of generating electricity from natural gas to be competitive with the price of electricity from coal. However, there are important environmental concerns related to fracking, including contamination of drinking water supplies, air quality pollution, and the production of GHG emissions (Navarro 2012; Myhrvold and Caldeira 2012).

Ironically, recent research demonstrates that increasing our use of natural gas in place of coal or oil—seen by some as a partial solution to global warming—may not be the solution it purports to be. This is because natural gas contains considerable amounts of methane, a far more powerful GHG than carbon dioxide. For this reason, even small amounts of gas venting and leaking lead to a very large GHG footprint (Howarth et al. 2012). However, scientific estimates of the amount of methane that leaks from wells, pipelines, and other facilities during natural gas production and delivery have varied considerably (Associated Press 2013).

Many states have set requirements, also known as standards, for expanding the use of energy efficiency and renewable energy. In addition, some states have set energy efficiency and/or renewable energy goals that are not legally binding. In

2013, twenty states have energy-efficiency resource standards and seven states have goals (North Carolina State University 2013a). Twenty-nine states, Washington D.C., and two US territories have renewable portfolio standards. In addition, eight states and two territories have renewable portfolio goals (North Carolina State University 2013b). These programs can have measurable impacts: for example, the World Resources Institute estimates that in 2007 California was second only to Texas in the United States in GHG emissions, and was the fourteenth largest emitter of such emissions globally (California Air Resources Board 2011, 3). Many of these programs set targets for 2020.

Planning for GHG emission reductions beyond 2020, a number of US states and Canadian provinces created North America 2050. The purpose of this partnership is to continue to make progress on "policies that move their jurisdictions toward a low-carbon economy while creating jobs, enhancing energy independence and security, protecting public health and the environment, and demonstrating climate leadership" (North America 2050 2012). In 2012, members included sixteen US states and four Canadian provinces.

In 2012, three states along the West Coast and British Columbia agreed to create the West Coast Infrastructure Exchange to develop new financing mechanisms for infrastructure updates. The agreement states that the exchange will facilitate investment of $1 trillion over the next 30 years and ensure the investments consider factors affecting infrastructure risks related to climate change (West Coast Infrastructure Exchange 2012).

Subnational cooperation on climate change is an important method for furthering climate-change mitigation and adaptation efforts and sharing information on best practices. For example, sixty subnational government leaders signed a statement during the UNFCCC conference of the parties in Copenhagen in 2009. The statement indicated that 50 to 80 percent of actions to reduce GHG emissions and adapt to climate change are carried out at the subnational level of government. In addition, the sixty subnational leaders committed to take specific actions at the subnational level, including planting 1 billion trees by 2015. The statement called on national governments to agree to ambitious targets to cut GHG emissions and recognize the complementary role of subnational governments in achieving the goals of the UNFCCC (UNFCCC 2009). In 2012, subnational leaders in the nonprofit Climate Group reported they have planted half of the trees needed to achieve their 2015 goal (Climate Group 2012b). The memorandum of understanding on subnational cooperation signed by the US State Department and China in 2011 is another example. This memorandum supports opportunities for state and local government leaders to share ideas and discuss best practices with Chinese provincial and local government leaders on topics such as trade, investment, energy, and the environment (US State Department 2011). In addition, the UNFCCC encourages communities to share strategies for climate adaptation through a database on local coping strategies (UNFCCC 2013a).

Politics of environmental policy at the state level can be different from those nationally. This is true for climate change: "Contrary to the kinds of political brawls so common in debates about climate change policy at national and international venues ... state-based policymaking has been far less visible and contentious, often cutting across traditional partisan and interest group fissures" (Rabe 2002, 22). Some legislation, like that involving motor vehicles in California, faces industry opposition but still is supported by several different groups—in the California case, environmentalists, Silicon Valley business leaders, and big cities (Rabe 2004, 143; Baldassare, Bonner, and Shrestha 2012, 17): American states continue to play an important role in reducing GHG emissions and adapting to climate change. "This pattern is also evident in many other nations around the world that allow for intergovernmental sharing of authority for energy, environmental protection and all other policy areas related to climate change" (Rabe 2011, 39).

At the state level, there also is room for what political scientists call *policy entrepreneurs* to move climate-change policy. Policy entrepreneurs know where change is possible and have expertise that is not seen as politically motivated: "These entrepreneurs have tailored policies to the political and economic realities of their particular setting and have built coalitions that seem almost unthinkable when weighed against the past decade of federal-level experience" (Rabe 2002, 151). The activity of the states may provide lessons for the federal government if the United States decides to change its formal policies. In the meantime, the states are learning from each other, and ideas are flowing back and forth through interstate meetings and conferences.

Cities are learning from each other as well. In 2005, the mayor of Seattle, Greg Nickels, began a program within the US Conference of Mayors to promote the goals of the Kyoto Protocol though actions by American cities. More than 1,000 mayors have signed the agreement. The Climate Protection Center provides information on best practices for climate action and issues annual awards for cities demonstrating leadership in climate protection (US Conference of Mayors Climate Protection Center 2009, 2012).

Participating cities commit to meet or exceed Kyoto Protocol targets of the first commitment period in their communities;

encourage the state and federal government to meet or exceed Kyoto goals; and urge the US Congress to pass GHG reduction legislation. The agreement's action plan discourages urban sprawl, promotes public transportation and the creation of car-pooling and bicycle lanes, and encourages the use of alternative energy sources. Environmentalists conclude the time is right for this group to encourage cities to achieve the goals of the second commitment period of the Kyoto Protocol: 18 percent below 1990 levels in the period from 2013 to 2020 (Doha Amendment 2012).

Megacities around the world, including ten in the United States, have created the C40 Climate Leadership Group to address climate change. The group addresses climate-change mitigation and adaptation. One in twelve people worldwide are represented by the C40 group. The C40 group provides information on over 4,500 actions by local governments to address climate change (C40 Climate Leadership Group 2011). The C40 group worked with Local Governments for Sustainability, the World Resources Institute, the World Bank, and programs at the United Nations to develop a global protocol for community-scale GHG emissions. A pilot version of the GHG protocol was released in May 2012. As of July 2013, thirty-three cities were testing the pilot version (World Resources Institute and World Business Council for Sustainable Development 2013).

California: Register, Clean, Renew

Climate change generates various responses—indifference, scorn, alarm, advocacy, and action. Climate change will affect families and economies in different ways. In some coastal areas, for example, it will have direct and negative impacts, while elsewhere the effects will be mild or beneficial.

California is working to address climate change on many levels, and many of its environmental initiatives have become national policy. The size of California's economy and population and the volume of its carbon dioxide emissions indicate that activities to reduce GHG emissions in this one state can influence efforts around the globe.

For more than twenty years, California has promoted energy efficiency, pollution reduction, and renewable energy. These are part of a *no-regrets policy* on climate change: "California's energy policies have been carried out primarily to reduce costs to consumers and air pollution. However, these 'no regrets' policies have set California on a firm path to respond effectively to growing concerns about the effects of human-caused greenhouse gas emissions on the earth's ecosystem" (California Energy Commission 1998).

In addition to standards for energy efficiency, pollution reduction, and renewable energy, California has pursued its goals to reduce GHG emissions through: 1) a climate action registry for voluntary reductions (Climate Action Reserve 2013); 2) a cap-and-trade system for mandatory reductions in the electricity, industrial, and transportation sectors (California Air Resources Board 2013); and 3) programs to increase the use of zero-emission vehicles in California (California Office of the Governor 2012). In California's cap-and-trade program, 25 percent of the funds from the sale of GHG allowances must be used to benefit disadvantaged communities in California, and 10 percent must fund projects in disadvantaged communities.

Much of California's current climate policy builds on executive orders during Governor Arnold Schwarzenegger's time in office. In early June 2005, Governor Schwarzenegger signed an executive order on climate change before an audience of business and environmental leaders at the United Nations World

Environmental Day conference in San Francisco. The governor declared, "As of today, California is going to be the leader in the fight against global warning. ... I say the debate is over. We know the science, we see the threat, and the time for action is now" (Bustillo 2005).

Under this executive order, the state will reduce its GHG emissions to 2000 levels by 2010 and to 1990 levels by 2020. In addition, it will reduce its GHG emissions to 80 percent below 1990 levels by 2050. The governor's plan included increasing energy efficiency and getting a third of California's electricity from renewable resources.

These actions are broadly supported by Californians. From July 2003 through July 2012, more than 70 percent of California residents believed immediate steps should be taken to counter the effects of global warming. Furthermore, in 2012, 77 percent supported requiring 33 percent renewable energy by 2020, although support was only 44 percent if achieving this goal meant higher electricity bills (Baldassare, Bonner, and Shrestha 2012, 14, 21).

New York

New York is also a leader in GHG emission-reduction policies. For example, the New York State Energy Research and Development Authority, the New York Power Authority, and the Long Island Power Authority have all invested billions of dollars in existing, expanded, and new energy-efficiency and renewable-energy programs (New York State Department of Environmental Conservation 2013). As of spring 2013, more than 110 communities in New York had pledged to become "Climate Smart." Climate Smart communities will encourage and support the development of "green" businesses and jobs in fields such as renewable energy, energy efficiency, or weatherization. The Climate Smart website displays methods and examples of GHG capture, local emissions reductions, and climate-adaptation strategies (New York State Department of Environmental Conservation 2013).

New York State has set two goals to help minimize climate change risk: reduce emissions of heat-trapping GHGs by 40 percent by 2030 and 80 percent by 2050 (relative to 1990 levels); and improve resilience to climate change in all communities throughout the state (2013). A 2009 executive order established GHG emission reduction goals and mandated the preparation of a Climate Action Plan.

In 2011, the New York State Energy Research and Development Authority published a report entitled "Responding to Climate Change in New York State," commonly referred to as the "ClimAID" report (Rosenzweig et al. 2011). The ClimAID report assessed the potential economic costs of climate-change effects and climate-change adaptation in eight major economic sectors in New York State, including agriculture, ecosystems, public health, water resources, ocean and coastal zones, energy, transportation, and communications. Unless adaptive measures are undertaken, the report found, climate change costs in New York State for the sectors listed could approach $10 billion annually by the middle of the century. However, the report states that a wide range of wisely chosen and timely adaptive strategies could markedly reduce climate change impacts in the state by amounts that exceed their costs, making them financially sound investments. Moreover, when noneconomic objectives such as ecosystems and social equity are taken into account, implementing adaptive strategies becomes even more cost-effective (Rosenzweig et al. 2011).

The New York State Climate Action Plan Interim Report, published in 2010, includes an inventory and forecast of greenhouse gases and identifies policy options for reducing GHG emissions and increasing climate resiliency. The report focuses on policy options that keep costs down, provide co-benefits for New York, and address other important issues, such as environmental justice. The report highlights the need for a portfolio of options to achieve the 2050 GHG emission-reduction goal, because more than one policy is needed to reach the goal. The report indicated that three policies have the largest potential to reduce GHG emission reductions in New York: high-penetration renewable energy policy; energy-efficiency incentives; and vehicle efficiency standards (New York State Climate Action Council 2010, OV-22).

The New York State Climate Action Plan Interim Report includes specific policy options to reduce climate vulnerability in the state. For example, the report recommends policies to integrate sea level rise and flood projections into government agency decisions; ensure critical infrastructure is resilient to extreme weather events; and continued monitoring, upkeep of warning systems, and research to address the agricultural and public health impacts of climate change (New York State Climate Action Council 2010, OV-37 to OV-48).

At the same time that New York is planning for drastic reductions in GHG emissions and preparing for possible increases in climate-related public health and infrastructure challenges, it is also embroiled in turmoil regarding hydraulic fracturing (fracking) by the oil and gas industry. New York, even more than other regions of the United States, faces major questions about the extent of its approval of fracking.

Beginning in 2008, fracking was temporarily prohibited in the state of New York while its environmental impacts were reviewed by the Department of Conservation. In 2011, the New York State Department of Environmental Conservation issued new recommendations for the oil and gas industry. Drilling is prohibited in and within buffer zones surrounding New York City and Syracuse watersheds and all primary aquifers; it is also prohibited on all state-owned land, including forested areas, wildlife management areas, and parks. Drilling will be permitted on privately held lands; however, it is subject to "rigorous and effective controls" (New York State Department of Environmental Conservation 2011). In 2012, New York Governor Cuomo began to consider allowing fracking to occur on a limited basis in parts of five counties near the Pennsylvania border, but only in communities that have expressed their support for fracking operations.

New Yorkers and coastal areas in neighboring states are taking steps to increase resilience and reduce the risk of damage from rising sea levels, storm surges, and heightened risk of severe storms. In January 2013, New York City waived height restrictions and raised the minimum elevation up to which buildings must be "flood proof." The flood proofing requirement applies to new construction and substantial alterations to existing buildings. According to the Mayor's Office, "New construction or buildings with substantial damage in need of repair must protect the structures by building at least one foot above the flood elevation currently required in the building code. The added elevation will provide a further margin of safety from potential flood damage, serve to enhance life safety and reduce property loss" (New York City Mayor's Office 2013). Also, in response to new flood-risk data from the Federal Emergency Management Agency, some homes flooded by Hurricane Sandy are being lifted five to ten feet to reduce the risk of flooding a second time and reduce the cost of flood insurance (Harris 2013).

In June 2013, seven months after Superstorm Sandy swamped lower Manhattan and coastal regions of outlying boroughs, New York City Mayor Michael Bloomberg announced the development of a plan to protect New York City from sea level rise and storm surges. The plan is based on new projections of climate impacts from a scientific panel convened by the city. By the 2050s, it is now believed, 800,000 people may be living within a flood zone covering a quarter of the city, which has 520 miles of coastline. Under the proposed plan, which is expected to cost approximately $20 billion, removable floodwalls would be constructed along the waterfront in lower Manhattan. Smaller barrier systems may also be built to defend waterways that can carry floodwaters inland. Levees, gates, and other defenses would be erected in specific critical locations around the city. In addition, marshes would be constructed to serve as buffers to rising sea waters, severe weather events, and storm surges. The panel's recommendations also include expanding beaches and constructing dunes along hard-hit coastlines in Staten Island and the Rockaways and building bulkheads in certain vulnerable neighborhoods in all the boroughs. In addition, a levee and a large new "Seaport City" are proposed for development in lower Manhattan. The plan also proposes giving over a billion dollars in grants to property owners to flood-proof their homes and to nursing homes and hospitals for flood-proofing and for upgrading pumps and electrical equipment to achieve high standards of flood protection. Mayor Bloomberg acknowledged that some of the proposals would be controversial. In Staten Island, for example, recommendations call for the construction of 15-to-20 foot levees to protect a section of the island. Barriers of this sort could have adverse impacts on ocean views and beach access (Peltz 2013).

Other Actions

The private sector has also emerged as a major player in the climate-change story in the United States. Many corporations have begun to reduce the financial costs of controls. The World Resources Institute in 2004 concluded that "proactive work" to measure emissions and minimize the costs of rule compliance could be much less costly than "reacting to events at a later date" (Fialka and Ball 2004; Metzger et al. 2012). Some utilities have taken early steps. New York's Consolidated Edison saved millions of dollars by eliminating natural gas leaks. And the company's GHG emissions in 2011 were more than 40 percent below its emissions for 2005 (ConEdison 2012). Johnson and Johnson plans to reduce GHG emissions 20 percent below 2010 levels by 2020 through energy efficiency and renewable energy (Johnson and Johnson 2013).

The US EPA provides resources and benchmarking tools through the Center for Corporate Climate Leadership. In 2012, ten corporations received awards recognizing excellence in achieving ambitious GHG emission reduction goals. International Paper, SC Johnson, and Intel Corporation are among the leaders recognized by this award. For GHG emission management of supply chain processes, 2012 awards went to the Port of Los Angeles, the United Parcel Service, and SAP (a software company based in Germany) (EPA Center for Corporate Climate Leadership 2013).

Individual changes in behavior have also been suggested to combat climate change. These are lessons of environmental education. While some efforts may seem unlikely and their effects insufficient, in aggregate, they can yield substantial emission reductions. For example, in general, more energy is required to produce packaging and newsprint from raw materials than re-cycled goods. Similarly, public transportation, carpooling, cy-cling, telecommuting, and flexible work schedules also reduce GHG emissions per person.

Rather than viewing prevention and mitigation of climate change as two separate areas, government, business, and agri-culture are beginning to link the two and to identify additional actions that could reduce risk to life and property. Without sizeable reductions of GHG emissions, efforts to address cli-mate change—especially adaptation of existing water controls and coastal infrastructure—will become more difficult and costly (Claussen 2004; Klein et al. 2007, 747, 759–760; Moser et al. 2012).

Groups like the Center for Climate and Energy Solutions, formerly known as the Pew Center on Global Climate Change, promote and advocate a large number of actions that can in-fluence future climate. Because some GHGs, particularly CO_2, may remain in the air for one hundred years or more, actions taken today to slow emissions will be most effective if they are designed to withstand changes in climate. Table 5.3 highlights a number of possible current actions that can reduce emissions and affect the amount and impact of climate change.

Over the past decade, rapid technological advancements and increased political pressure have led to lower costs for re-newable-energy options. Many European countries, including Denmark and Germany, and many US states have established programs for increasing the use of renewable energy. Large hydro-electricity facilities provide about 10 to 20 percent of California's electricity, depending on the weather. In addition, about 15 percent of California's energy supplies are currently provided by geothermal, biomass, small hydropower, solar, and wind renewable sources (California Energy Commission 2012). Excluding large hydropower, by 2020 California plans to use renewable energy for 33 percent of electricity sales. Also, California is working to advance zero-emission vehicles and has a goal of 1.5 million zero-emission vehicles by 2025.

According to a report released in 2013 by the IEA (2013b), power generation from renewables will increase by 40 percent worldwide within the next five years, with developing nations responsible for two-thirds of the projected growth. Renewables are becoming popular in developing countries such as China, where they help to address and diversify a rapidly increasing demand for power while both minimizing the pollution associated with energy production and mitigating climate change. On a global basis, hydropower, wind, solar, and other renewable energy sources are expected to generate more energy than natural gas by 2016. Renewables are also expected to be responsible for generating double the energy of nuclear plants by 2016. The use of biofuels in the transportation industry and for heat production will also increase, but the growth rate for biofuel energy is expected to be slower than for other types of renewable energy. However, in spite of a promising future for renewables, the IEA report cautions that the industry may face political challenges as citizens and policy makers in developed countries with struggling economies challenge the costs of incentive programs that support the research, development, installation, and use of these technologies (IEA 2013b).

There is considerable room for change worldwide, where nonrenewable fossil fuels remain more than 80 percent of the supply of primary energy. In 2010, more than 16 percent of final energy consumption came from renewable resources; growth in renewable energy is strong (Renewable Energy Policy

Table 5.3
Current Actions That Can Influence Future Climate Changes

Activity	Possible Impact of Climate Change on Activity	Prevention	Actions Needed if Impact not Prevented
Carbon sequestration in forests	Insect infestation of forests due to drought	Manage forests for drought conditions using sustainable forest-stewardship methods.	Conduct fire-prevention procedures and sustainable biopower generation to contain greenhouse gas release from fire-prone dead wood.
Energy efficiency (voluntary/short-term)	Temperature swings reducing the willingness or ability of consumers to conserve energy	Follow conservative building practices, such as window and building orientation, deciduous shading, insulation, and flow-through air circulation.	Provide increased incentives for consumers to conserve during periods of peak demand.
Renewable energy	Hydroelectricity reduced, wind more extreme, biomass uncertain, solar uncertain (more clouds)	Accelerate renewable energy development to help reduce greenhouse gas emissions.	Conduct research and development to improve operations over a broad range of weather conditions.
Agriculture	Droughts or inundation leading to changes in the timing or type of water problems	Employ "no-till" agricultural techniques. Expand the use of waste methane for energy generation.	Change the type of crops to better reflect timing and availability of water supply.

Activity	Possible Impact of Climate Change on Activity	Prevention	Actions Needed if Impact not Prevented
Transportation	Hazardous driving conditions becoming more frequent (e.g., limited visibility, limited traction, extreme heat)	Expand best practices, such as "smart growth," to reduce sprawl and promote public transportation, telecommuting, alternative fuels, and better fuel efficiency.	Provide automakers with incentives for improving vehicle performance under hazardous conditions.
Land use: coastal	Rising sea levels and severe weather leading to greater erosion and mud slides	Create incentives for setbacks from coastal areas and bluffs. Build and strengthen sea walls and levies. Create areas of high ground where possible.	After severe storms, evoke eminent domain with due compensation, and prevent reconstruction within setback area from coast.
Land use: flood plain	Less snow and more rain leading to greater flood risk	Change the flow-through dams with new patterns. Build and expand cisterns, floodplain setbacks, and levies to slow and direct floodwaters. Separate storm from sewage drains.	After severe storms, evoke eminent domain with just compensation, and prevent reconstruction within setback area from rivers.
Land use: mountain	Extreme weather events leading to greater erosion risk	Maintain healthy vegetation on slopes to help control erosion.	Increase reservoir dredging to maintain the water-storage capacity of dams.

Network for the 21st Century 2012). However, the global use of coal is growing as well, even though coal-fired energy is decreasing in some countries (IEA 2013a). Compared to electricity generated by a conventional pulverized coal plant, wind- and solar-generated electricity produces from 130 to 268 times less CO_2 per kilowatt hour (Pirages and Cousins 2005).

Conclusion

Through international organizations, national governments, businesses, and other interest groups, people are learning about climate change and are responding to this extraordinary international challenge. Despite political activity in many nations that obfuscates the cost of inaction and does little to discourage delay, international efforts have been undertaken, innovations in technologies and policies have been widespread, and some nations, states, and provinces now require GHG emission reductions. Recognizing that some amount of climate change has already been set in motion, many communities and states have developed plans for climate adaptation.

Through international initiatives and through focused and sometimes dramatic journalistic coverage of the causes and effects of climate change, global learning on climate change is ongoing. Some scientists have concluded that the present response is not enough, but the world community or at least a large portion of it has proved to be capable of reacting to environmental challenges quickly when the true problem is made clear. On the other hand, as Emanuel points out:

Scientists, engineers, and economists can do no more than formulate options for dealing with the risks. It is up to society as a whole to decide what combination of options to deploy. This is a terrifically difficult decision because the costs may be high and those paying them are not likely to be serious beneficiaries of their own actions. Indeed, there are few, if any, historical examples of civilizations consciously making sacrifices on behalf of descendants two or more generations removed. (2012, 76)

Still, some progress is being made. As a top-down international effort goes forward, so do activities on the state and provincial levels, in businesses, and at the individual level. Experiments are ongoing in Denmark, Germany, Japan, the United Kingdom, California, and New Jersey; at corporations committed to GHG-emission reductions; and in neighborhoods and cities throughout the world. As doubts about the existence of climate change and the human contribution to it are removed, efforts to slow it down and reduce vulnerability to its effects have accelerated.

References

ABCNews.com. 2001. Poll, April 11–15. http://www.pollingreport.com/enviro3.htm, accessed September 18, 2013.

Anguelovski, Isabelle, and Debra Roberts. 2011. Spatial justice and climate change: Multiscale impacts and local development in Durban, South Africa. In JoAnn Carmin and Julian Agyeman, edited by *Environmental Inequalities Beyond Borders: Local Perspectives on Global Injustices*. Cambridge, MA: MIT Press.

Associated Press. 2013. EPA lowered estimates of methane leaks during natural gas production. April 28. http://fuelfix.com/blog/2013/04/28/epa-lowered-estimates-of-methane-leaks-during-natural-gas-production, accessed July 8, 2013.

Baldassare, Mark, Sonja Bonner, and Petek Jui Shrestha. July 2012. Californians and the Environment. Public Policy Institute of California survey. http://www.ppic.org/content/pubs/survey/S_712MBS.pdf, accessed July 8, 2013.

Borenstein, Seth. 2012. Global warming talk heats up, revisits carbon tax. *The Seattle Times*. November 15. http://seattletimes.com/html/businesstechnology/2019673844_apusscisuperstormclimatetalk.html, accessed July 8, 2013.

Boxall, Bettina. 2006. Bush's grade on environment falls. *Los Angeles Times*. August 4, A-22.

British Columbia Ministry of Finance. 2012. Budget and Fiscal Plan 2012/13–2014/15. February 21. http://www.bcbudget.gov .bc.ca/2012/bfp/2012_Budget_Fiscal_Plan.pdf, accessed September 18, 2013.

Broder, John. M. 2010. "Cap and trade" loses its standing as energy policy of choice. *New York Times*, New York edition. March 26, A13.

Bruntland, G., ed. 1987. *Our Common Future: The World Commission on Environment and Development*. Oxford: Oxford University Press.

Bryner, Gary, and Robert J. Duffy. 2012. *Integrating Climate, Energy, and Air Pollution Policies*. Cambridge, MA: MIT Press.

Burger, Andrew. 2012. Solar energy system approvals surge following launch of Japan's renewable energy feed-in tariff. Clean Technica. October 12. http://cleantechnica.com/2012/10/12/solar-energy-systems-approvals-surge-following-launch-of-japans-renewable-energy-feed-in-tariff, accessed February 17, 2013.

Burwell, Sylvia Mathews and John P. Holdren. 2013. Science and technology priorities for the FY 2015 budget. Memorandum for the heads of executive departments and agencies. July 26. http://www.whitehouse.gov/sites/default/files/microsites/ostp/fy_15_memo_m-13-16 .pdf, accessed September 23, 2013.

Bustillo, Miguel. 2005. A shift to green. *Los Angeles Times*. June 12, C1, C5.

C40 Climate Leadership Group. 2011. C40 Cities. http://www .c40cities.org/c40cities.

California Air Resources Board. 2011. California Greenhouse Gas Emissions Inventory: 2000–2009. December. http://www.arb.ca.gov/ cc/inventory/pubs/reports/ghg_inventory_00-09_report.pdf, accessed July 8, 2013.

California Air Resources Board. 2013. Cap and Trade Program. Last reviewed June 24. http://www.arb.ca.gov/cc/capandtrade/ capandtrade.htm, accessed July 8, 2013.

California Energy Commission. 1998. *Staff Report*, vol. 1. Executive Summary 1997 Global Climate Change Greenhouse Gas Emissions

Reduction Strategies for California. Sacramento, CA: California Energy Commission.

California Energy Commission. 2012. 2011 total system power in gigawatt hours. http://energyalmanac.ca.gov/electricity/total_system_power.html.

California Office of the Governor. 2012. Executive order B-16–2012. http://www.gov.ca.gov/news.php?id=17472.

Carbon Tax Center. 2012. Archive for the "international" category. http://www.carbontax.org/blogarchives/category/international.

Carpenter, Chad, Pamela Chasek, Peter Doran, Emily Gardner, and Daniel Putterman. 1996. Summary report of the Second Conference of the Parties to the Framework Convention on Climate Change, 8-19 July 1996. *Earth Negotiations Bulletin.* International Institute for Sustainable Development. 12 (38) Monday 22 July 1996. http://www.iisd.ca/download/pdf/enb1238e.pdf, accessed September 18, 2013.

Carpenter, Chad, Pamela Chasek, and Steve Wise. 1995. Summary of the First Conference of the Parties for the Framework Convention on Climate Change, 28 March–17 April, 1995. *Earth Negotiations Bulletin.* International Institute for Sustainable Development. 12 (21) Monday, 10 April 1995. http://www.iisd.ca/download/pdf/enb1221e.pdf, accessed September 18, 2013.

Center for Climate and Energy Solutions. 2011. Climate Change 101: State Action. January. http://www.c2es.org/docUploads/climate101-state.pdf.

Chang, Shiyan, Lili Zhao, Govinda R. Timilsina, and Xiliang Zhang. 2012. Development of biofuels in China: Technologies, economics and policies. Report No. WPS6243 Policy Research Working Paper.

Claussen, Eileen. 2004. An effective approach to climate change. *Science* 306 (5697):816.

Climate Action Reserve. 2013. Climate Action Reserve monthly newsletter. July. http://www.climateactionreserve.org/wp-content/uploads/2009/03/2013.06-July-Climate-Action-News.pdf, accessed July 8, 2013.

Climate Group. 2012a. States and regions. http://www.theclimategroup.org/programs/states-and-regions, accessed September 18, 2013.

Climate Group. 2012b. Sub-national governments half way to 1 billion trees target. June 2. http://www.theclimategroup.org/what-we-do/news-and-blogs/sub-national-governments-half-way-to-1-billion-trees-target, accessed September 1, 2013.

ConEdison. 2012. 2011 Sustainability report: Reducing greenhouse gases. http://www.conedison.com/ehs/2011annualreport/environmental-stewardship/reducing-greenhouse-gases/air-emissions-performance.html, accessed September 18, 2013.

Conference of New England Governors and Eastern Canadian Premiers. 2001. Climate Change Action Plan. August. http://www.gov.ns.ca/nse/climate.change/docs/NEG-ECP.pdf, accessed July 8, 2013.

Davidson, Osha Gray. 2012. So far so good for Germany's nuclear phase-out, despite dire predictions. Clean Break: Chapter 4 in the story of Germany's switch to renewables. November 16. *Inside Climate News.* http://insideclimatenews.org/print/22509.

Deutsche Welle. 2007. EU pressures US to reach climate consensus in Bali. December 13. http://www.dw.de/eu-pressures-us-to-reach-climate-consensus-in-bali/a-3002780, accessed July 8, 2013.

Doha Amendment. 2012. Doha amendment to the Kyoto Protocol to the United Nations Framework Convention on Climate Change. Doha. December 8. Not yet in force as of September 2013. http://unfccc.int/kyoto_protocol/doha_amendment/items/7362.php, accessed September 22, 2013.

Dolsak, Nives. 2001. Mitigating global climate change: Why are some countries more committed than others? *Policy Studies Journal: the Journal of the Policy Studies Organization* 29 (3):414–436.

Emanuel, Kerry. 2012. *What We Know About Climate Change.* 2nd ed. Cambridge, MA: MIT Press.

ENE. 2013. New Jersey: emissions and emissions reduction targets. Appendix and state profiles. Climate Vision 2020. Environment Northeast. http://www.eneclimatevision.org/appendix-state-profiles/new-jersey, accessed September 22, 2013.

Environmental Protection Agency. 2009. Endangerment and Cause or Contribute findings for greenhouse gases under the Clean Air Act. Federal Register. http://www.epa.gov/climatechange/Downloads/endangerment/Federal_Register-EPA-HQ-OAR-2009-0171-Dec.15-09.pdf.

Environmental Protection Agency. 2011. Program announcement: Adoption of an energy efficiency design index for international shipping. July. http://www.epa.gov/otaq/regs/nonroad/marine/ci/420f11025.pdf.

Environmental Protection Agency. 2012a. Greenhouse Gas Reporting Program. GHGRP 2010: Reported Data. http://www.epa.gov/ghgreporting/ghgdata/reported-2010/index.html, accessed September 18, 2013.

Environmental Protection Agency. 2012b. International Climate Partnerships. http://epa.gov/climatechange/EPAactivities/internationalpartnerships.html, accessed September 1, 2013.

Environmental Protection Agency Center for Corporate Climate Leadership. 2013. 2012 Climate Leadership Award winners. http://www.epa.gov/climateleadership/awards/2012winners.html, accessed June 30, 2013.

EurActiv. 2013. Doha climate change deal clears way for "damage aid" to poor nations. EurActiv.com. Updated January 8. http://www.euractiv.com/climate-environment/doha-climate-change-deal-clears-news-516547, accessed September 17, 2013.

European Commission. 2003. Energy: Issues, options and technologies: A survey of public opinion in Europe. Press release. March 6. http://ec.europa.eu/research/press/2003/pr0603en.html, accessed July 8, 2013.

European Commission. 2012. Report from the Commission to the European Parliament and the Council: Progress towards achieving the Kyoto objectives. October 24. http://eur-lex.europa.eu/LexUriServ/LexUriServ.do?uri=COM:2012:0626:FIN:EN:PDF.

European Commission. 2013a. The EU Emissions Trading System. http://ec.europa.eu/clima/policies/ets/index_en.htm.

European Commission. 2013b. What is the EU doing about climate change? January 7. http://ec.europa.eu/clima/policies/brief/eu/index_en.htm.

Federal Republic of Germany. 2010. Energy concept for an environmentally sound, reliable, and affordable energy supply. Federal Ministry of Economics and Technology and Federal Ministry for the Environment, Nature Conservation and Nuclear Safety. http://www.bmu.de/fileadmin/bmu-import/files/english/pdf/application/pdf/energiekonzept_bundesregierung_en.pdf, accessed July 8, 2013.

Fialka, John J., and Jeffrey Ball. 2004. Companies get ready for greenhouse gas limits. *Wall Street Journal.* October 26, A2.

Gallup. 2003. No environmental backlash against Bush administration though most Americans favor strong environmental policies. Press release, April 21. Gallup poll on the environment conducted March 3–5, 2003. http://www.gallup.com/poll/8215/Environmental-Backlash-Against-Bush-Administration.aspx, accessed July 8, 2013.

Gallup. 2012. Americans endorse various energy, environment proposals: Republicans and Democrats show substantially differing levels of support. Press release, April 9. Gallup poll on the environment conducted March 8–11, 2012. http://www.gallup.com/poll/153803/Americans-Endorse-Various-Energy-Environment-Proposals.aspx, accessed July 8, 2013.

Garnier, Donatien, and Cédric Faimali. 2010. United States: Gulf Coast, farewell to the Big Easy. In *Climate Refugees,* 164–199. Cambridge, MA: MIT Press.

Giddens, Anthony. 2011. *The Politics of Climate Change.* 2nd ed. Cambridge, UK: Polity Press.

Glanz, James. 2012. The cloud factories: Power, pollution and the Internet. *New York Times.* September 22.

Global Climate Coalition. n.d. Home page. http://www.globalclimate.org, accessed September 19, 2004 (no longer active as of March 2006).

Global Environment Facility. 2012a. China: Mainstreaming adaptation to climate change into water resources management and rural development. Project executive summary. http://www.thegef.org/gef/project_highlights/SCCF/China, accessed September 1, 2013.

Global Environment Facility. 2012b. Mainstreaming adaptation to climate change into water resources management and rural development. Project identification form. http://www.thegef.org/gef/project_detail?projID=3265, accessed September 1, 2013.

Goreham, Steve. 2013. Letter from S. Fred Singer, March 2013. Included with mailing of *The Mad, Mad, Mad World of Climatism: Mankind and Climate Change Mania.* New Lenox, IL: New Lenox Books.

Government of Japan. 2002. Japan's Third National Communication under the United Nations Framework Convention on Climate Change. http://unfccc.int/national_reports/annex_i_natcom/submitted_natcom/items/1395.php, accessed July 8, 2013.

Government of the Netherlands. 2013a. Agriculture and live-stock. http://www.government.nl/issues/agriculture-and-livestock/agriculture-and-horticulture, accessed April 27, 2013.

Government of the Netherlands. 2013b. Dutch greenhouse industry goes green. Ministry of Economic Affairs. http://www.hollandtrade.com/media/news/?bstnum=4057, accessed July 6, 2013.

Grose, Thomas K. 2013. Europe's carbon market crisis: Why does it matter? *National Geographic News.* April 18. http://news.nationalgeographic.com/news/energy/2013/04/130418-europe-carbon-market-crisis/.

Harris, Elizabeth A. 2013. Going up a few feet, and hoping to avoid a storm's path. *New York Times.* April 15. http://www.nytimes.com/2013/04/16/nyregion/after-hurricane-sandy-homeowners-elevate-property.html?pagewanted=all&pagewanted=print, accessed June 27, 2013.

Harris Interactive. 2002. Majorities continue to believe in global warming and support Kyoto Treaty: Poll, September 19–23. http://www.harrisinteractive.com/NEWS/allnewsbydate.asp?NewsID=535, accessed July 8, 2013.

Hattori, Takashi. 2007. The rise of Japanese climate change policy: Balancing the norms of economic growth, energy efficiency, international contribution, and environmental protection. In *The Social Construction of Climate Change: Power, Knowledge, Norms, Discourses,* edited by Mary E. Pattenger, 75–97. Surrey, UK: Ashgate.

Heggelund, Gorild, Steinar Andresen, and Inga Fritzen Buan. 2010. Chinese climate policy: Domestic priorities, foreign policy, and emerging implementation. In *Global Commons, Domestic Decisions,* edited by Kathryn Harrison and Lisa McIntosh Sundstrom, 239–261. Cambridge, MA: MIT Press.

Holdren, John P. 2013. Statement of Dr. John P. Holdren Director, Office of Science and Technology Policy Executive Office of the President of the United States to the Committee on Science, Space and Technology United States House of Representatives on Research and Development in the President's Fiscal Year 2014 Budget. April 17. http://science.house.gov/sites/republicans.science.house.gov/files/documents/HHRG-113-SY-WState-JHoldren-20130417.pdf, accessed September 23, 2013.

Howarth, Robert, Drew Shindell, Renee Santoro, Anthony Ingraffea, Nathan Phillips, and Amy Townsend-Small. 2012. Methane emissions from natural gas systems. Background paper, National Climate Assessment Reference No. 2011–0003. February 25.

Intergovernmental Panel on Climate Change. 1990. First Assessment Report Overview and Policy Maker Summaries and 1992 IPCC Supplement. Report. Geneva, Switzerland: IPCC. http://www.ipcc .ch/publications_and_data/publications_ipcc_90_92_assessments_far .shtml, accessed September 19, 2013.

Intergovernmental Panel on Climate Change. 1995. IPCC Second Assessment Report: Climate Change. Geneva, Switzerland: IPCC.

International Energy Agency. 2012. CO_2 Emissions from Fuel Combustion: 2012 highlights (pre-release). October 19. http://www.iea .org/publications/freepublications/publication/name,4010,en.html, accessed January 2, 2013.

International Energy Agency. 2013a. FAQs: Coal. http://www.iea.org/ aboutus/faqs/coal/, accessed July 5, 2013.

International Energy Agency. 2013b. Renewables to surpass gas by 2016 in the global power mix: IEA report sees renewable power increasingly cost-competitive with new fossil-fuel generation, but agency warns against complacency. Press Release. June 26. http://www .iea.org/newsroomandevents/pressreleases/2013/june/name,39156, en.html, accessed September 19, 2013.

Johnson & Johnson. 2013. Climate change. http://www.jnj.com/ responsibility/Environment/Climate-Change, accessed September 17, 2013.

Kameyama, Yasuko. 2012. Japan. Unpublished manuscript. November 10. Dr. Yasuko Kameyama, Head, Sustainable Social Systems Section, Center for Social and Environmental Systems Research, National Institute for Environmental Studies. Tsukuba, Japan.

Kameyama, Yasuko. 2013. Personal communication. Dr. Yasuko Kameyama, Head, Sustainable Social Systems Section, Center for Social and Environmental Systems Research, National Institute for Environmental Studies. Tsukuba, Japan. September 15, 2013.

Klein, R. J. T., S. Huq, F. Denton, T. E. Downing, R. G. Richels, J. B. Robinson, and F. L. Toth. 2007. Inter-relationships between adaptation and mitigation. In *Climate Change 2007: Impacts, Adaptation and Vulnerability. Contribution of Working Group II to the*

Fourth Assessment Report of the Intergovernmental Panel on Climate Change, edited by M. L. Parry, O. F. Canziani, J. P. Palutikof, P. J. van der Linden, and C. E. Hanson, 745–777. Cambridge, UK: Cambridge University Press.

Leiserowitz, A., E. Maibach, C. Roser-Renouf, G. Feinberg, J. Marlon, and P. Howe. 2013. Public Support for Climate and Energy Policies in April 2013. Yale University and George Mason University. New Haven, CT: Yale Project on Climate Change Communication.

Lesser, Adam. 2012. EBay's bet on fuel cells will influence data centers. *Bloomberg Businessweek*. October 31. http://www.businessweek.com/printer/articles/79092-ebays-bet-on-fuel-cells-will-influence-data-centers.

Lewis, Joanna. 2011. Energy and climate goals of China's 12th five-year plan. Center for Climate and Energy Solutions. http://www.c2es.org/international/key-country-policies/china/energy-climate-goals-twelfth-five-year-plan.

Maibach, Edward, Connie Roser-Renouf, Emily Vraga, Brittany Bloodhart, Ashley Anderson, Neil Stenhouse, and Anthony Leiserowitz. 2013. A national survey of Republicans and Republican-leaning independents on energy and climate change. George Mason University Center for Climate Change Communication and Yale Project on Climate Change Communication. http://environment.yale.edu/climate-communication/files/Republican_Views_on_Climate_Change.pdf.

Meckling, Jonas. 2011. *Carbon Coalitions: Business, Climate Politics, and the Rise of Emissions Trading*. Cambridge, MA: The MIT Press.

Metzger, Eliot, Samantha Putt Del Pino, Sally Prowitt, Jenna Goodward, and Alexander Perera. 2012. sSWOT: Sustainability Strengths, Weaknesses, Opportunities, Threats. World Resources Institute. http://pdf.wri.org/sustainability_swot_user_guide.pdf.

Moser, Susanne, Julia Ekstrom, and Guido Franco. 2012. *Our Changing Climate 2012*. Sacramento, CA: California Energy Commission.

Myhrvold, N. P., and K. Caldeira. 2012. Greenhouse gases, climate change and the transition from coal to low-carbon electricity. *Environmental Research Letters* 7 (17):014019. doi:10.1088/1748-9326/7/1/014019.

National Bureau of Statistics, China. 2011. Communiqué on the achievements of energy conservation targets by region in "11th Five-year Plan" period. June 17. http://www.stats.gov.cn/english/newsandcomingevents/t20110617_402732886.htm.

National Development and Reform Commission, China. 2007. China's national climate change programme. June. http://en.ndrc.gov.cn/newsrelease/P020070604561191006823.pdf, accessed July 8, 2013.

Navarro, Mireya. 2012. Bloomberg backs "responsible" extraction of gas and pays to help set up rules. *New York Times*. August 24. http://www.nytimes.com/2012/08/25/nyregion/bloomberg-backs-gas-drilling-with-rules-to-protect-the-environment.html, accessed September 1, 2013.

New York City Mayor's Office. 2013. Mayor Bloomberg announces new measures to allow home and property owners rebuilding after Hurricane Sandy to meet updated flood standards. January 31, No. PR-044–13. http://www.nyc.gov/cgi-bin/misc/pfprinter.cgi?action=print&sitename=OM&p=136Energy6517490000, accessed June 26, 2013.

New York State Climate Action Council. 2010. New York State Climate Action Plan Interim Report. http://www.dec.ny.gov/energy/80930.html, accessed September 19, 2013.

New York State Department of Environmental Conservation. 2011. New recommendations issued in hydraulic fracturing review. http://www.dec.ny.gov/press/75403.html, accessed July 8, 2013.

New York State Department of Environmental Conservation. 2013. Energy and climate: The energy/climate change connection. http://www.dec.ny.gov/60.html, accessed April 15, 2013.

Nicola, Stefan. 2013. Germany's greenhouse gas output rose in 2012 as coal use surged. Bloomberg. February 25. http://www.bloomberg.com/news/2013-02-25/germany-s-greenhouse-gas-output-rose-in-2012-as-coal-use-surged.html, accessed September 19, 2013.

North America 2050. 2012. Home page. http://na2050.org, accessed May 22, 2013.

North Carolina State University. 2013a. Database of State Incentives for Renewables and Efficiency: Energy Efficiency Resource Standards. February. http://www.dsireusa.org/documents/summarymaps/EERS_map.pdf.

North Carolina State University. 2013b. Database of State Incentives for Renewables and Efficiency: Renewable Portfolio Standard Policies. March. http://www.dsireusa.org/documents/summarymaps/RPS_map.pdf.

Osofsky, Hari M., and Lesley K. McAllister. 2012. *Climate Change Law and Policy*. New York: Wolters Kluwer Law & Business.

PBL Netherlands Environmental Assessment Agency. 2012. Trends in global CO_2 emissions. July 17. http://www.pbl.nl/en/publications/2012/trends-in-global-co2-emissions-2012-report, accessed December 6, 2013.

Peltz, Jennifer. 2013. Climate change in NYC: Mayor Bloomberg to discuss prepping city for warming world. *Huffington Post*. June 11. http://www.huffingtonpost.com/2013/06/11/climate-change-in-new-york-city-mayor-bloomberg_n_3420707.html, accessed June 27, 2013.

Pew Charitable Trusts. 2013. Who's winning the clean energy race? 2012 edition. Prepared by Bloomberg New Energy Finance for Pew Charitable Trusts. April. http://www.pewtrusts.org/uploadedFiles/wwwpewtrustsorg/News/Press_Releases/Clean_Energy/clen-G20-report-2012-FINAL.pdf, accessed July 8, 2013.

Pirages, Dennis, and Ken Cousins. 2005. *From Resource Scarcity to Ecological Security: Exploring New Limits to Growth*. Cambridge, MA: MIT Press.

Qiu, Jane. 2012. Chinese survey reveals widespread coastal pollution. *Nature*. November 6. http://www.nature.com/news/chinese-survey-reveals-widespread-coastal-pollution-1.11743, accessed September 1, 2013.

Rabe, Barry G. 2002. Greenhouse and statehouse: The evolving state government role in climate change. Washington, DC: Pew Center on Global Climate Change. http://www.c2es.org/docUploads/states_greenhouse.pdf, accessed July 8, 2013.

Rabe, Barry G. 2004. *Statehouse and Greenhouse: The Emerging Politics of American Climate Change Policy*. Washington, DC: Brookings Institute Press.

Rabe, Barry. 2011. Contested federalism and American climate policy. Paper presented at American Political Science Association Annual Meeting. http://papers.ssrn.com/sol3/papers.cfm?abstract_id=1902998, accessed September 1, 2013.

Renewable Energy Policy Network for the 21st Century. 2012. Renewables 2012 Global Status Report. http://www.ren21.net/Resources/Publications/REN21Publications/Renewables2012GlobalStatus-Report.aspx, accessed September 20, 2013.

Reuters. 2013. UPDATE 1: EU carbon hits new record low after back-loading vote. April 16. http://www.reuters.com/article/2013/04/16/eu-ets-vote-idUSL5N0D32PU20130416, accessed September 1, 2013.

Rosenzweig, C., W. Solecki, A. DeGaetano, M. O'Grady, S. Hassol, and P. Grabhorn, eds. 2011. Responding to climate change in New York State: The ClimAID integrated assessment for effective climate change adaptation. Prepared for the New York State Energy Research and Development Authority by Columbia University, the City University of New York and Cornell University. Albany, NY: New York State Energy Research and Development Authority. http://www.nyserda.ny.gov/Publications/Research-and-Development-Technical-Reports/Environmental-Reports/EMEP-Publications/Response-to-Climate-Change-in-New-York.aspx, accessed September 18, 2013.

Saad, Lydia. 2013. Americans' concerns about global warming on the rise: Majority believe global warming is happening, but many still say it's exaggerated. Gallup Politics, April 8. http://www.gallup.com/poll/161645/americans-concerns-global-warming-rise.aspx, accessed September 20, 2013.

Schreurs, Miranda A., and Yves Tiberghien. 2010. European Union leadership in climate change: Mitigation through multilevel reinforcement. In *Global Commons and Domestic Decisions*, edited by Kathryn Harrison and Lisa McIntosh Sundstrom. Cambridge, MA: MIT Press.

Scientific Committee on Oceanic Research. 2008. The ocean in a high-CO_2 world: Ocean acidification. Second Symposium, Monaco Declaration. http://www.ocean-acidification.net/Symposium2008/MonacoDeclaration.pdf, accessed July 8, 2013.

Scientific Committee on Oceanic Research. 2012. The ocean in a high-CO_2 world: Ocean acidification. Third Symposium, Program Information. http://www.highco2-iii.org/main.cfm?cid=2259&nid=14765.

SourceWatch. n.d. Global climate coalition. Last modified January 12, 2012. http://www.sourcewatch.org/index.php?title=Global_Climate_Coalition, accessed July 8, 2013.

St. Arnaud, Bill. 2012. Commentary: Using ICT for adaptation rather than mitigation to climate change. International Institute for Sustainable Development. October. http://www.iisd.org/pdf/2012/com_icts_starnaud.pdf.

Sussman, Dalia. 2001. Poll: Most Support Global Warming Treaty." *ABC News*. April 17, 2001. http://abcnews.go.com/US/story?id=93545&page=1, accessed September 17, 2013.

Tabuchi, Hiroko. 2012. Japan sets policy to phase out nuclear power plants by 2040. *New York Times*. September 14. http://www.nytimes.com/2012/09/15/world/asia/japan-will-try-to-halt-nuclear-power-by-the-end-of-the-2030s.html?pagewanted=all&pagewanted=print.

Tiberghien, Yves, and Miranda A. Schreurs. 2010. Climate leadership, Japanese style: Embedded symbolism and post-2001 Kyoto Protocol politics. In *Global Commons and Domestic Decisions*, edited by Kathryn Harrison and Lisa McIntosh Sundstrom. Cambridge, MA: MIT Press.

Tsukimori, Osamu. 2012. UPDATE 2: Tepco seeks more govt support as Fukushima costs soar. Reuters. November 7. http://www.reuters.com/article/2012/11/07/tepco-fukushima-idUSL3E8M77K720121107.

Union of Concerned Scientists. 2012. Ripe for retirement: The case for closing America's costliest coal plants. November. http://www.ucsusa.org/assets/documents/clean_energy/Ripe-for-Retirement-Executive-Summary.pdf.

United Kingdom Committee on Climate Change. 2013. Reducing the UK's carbon footprint and managing competitiveness risks. April. http://www.theccc.org.uk/wp-content/uploads/2013/04/CF-C_Summary-Rep_Bookpdf.pdf.

United Nations. 1992. Framework Convention on Climate Change. http:/unfccc.int, accessed September 18, 2013.

United Nations Environment Programme. 2013. CDM projects by host region. http://cdmpipeline.org/cdm-projects-region.htm#1, accessed July 5, 2013.

United Nations Framework Convention on Climate Change. 2009. Copenhagen Climate Change Conference. http://unfccc.int/meetings/copenhagen_dec_2009/meeting/6295.php, accessed November 11, 2012.

United Nations Framework Convention on Climate Change. 2012. Now, up to and beyond 2012: The Bali road map. http://unfccc.int/key_steps/bali_road_map/items/6072.php, accessed April 27, 2013.

United Nations Framework Convention on Climate Change. 2013a. Database on Local Coping Strategies. http://maindb.unfccc.int/public/adaptation, accessed September 20, 2013.

United Nations Framework Convention on Climate Change. 2013b. Decisions adopted by the Conference of the Parties. Report of the Conference of the Parties on its eighteenth session, held in Doha from 26 November to 8 December 2012. Addendum Part Two: Action taken by the Conference of the Parties at its eighteenth session. http:// unfccc.int/resource/docs/2012/cop18/eng/08a01.pdf, accessed October 30, 2013.

United Nations Framework Convention on Climate Change. 2013c. Meetings. http://unfccc.int/meetings/items/6240.php, accessed September 20, 2013.

United Nations Framework Convention on Climate Change. 2013d. Parties to the Convention and observer states. http://unfccc.int/ parties_and_observers/parties/items/2352.php, accessed September 20, 2013.

United Nations Framework Convention on Climate Change Conference of the Parties. 1996. Report of the conference of the parties on its second session, held at Geneva from 8 to 19 July 1996. Addendum. Part two: action taken by the conference of the parties at its second session. http://unfccc.int/resource/docs/cop2/15a01.pdf, accessed September 18, 2013.

United States Conference of Mayors Climate Protection Center. 2009. Mayors leading the way on climate protection. http://www.usmayors .org/climateprotection/revised, accessed July 8, 2013.

United States Conference of Mayors Climate Protection Center. 2012. Grand Rapids (MI) and Beaverton (OR) win first place honors for local climate protection efforts. June 13. http://usmayors.org/ pressreleases/uploads/2012/0614-climateprotectionawards.pdf, accessed August 19, 2012.

United States Global Change Research Program. 2012. Press Release: Administration Releases 10-Year Global Change Strategic Plan. April 27. http://downloads.globalchange.gov/strategic-plan/strategic_plan_ press_release.pdf, accessed September 20, 2013.

United States Global Change Research Program. 2013. National Institutes of Health explore impact of climate change on human health. April 22. http://globalchange.gov/whats-new/health-news/913-national-institutes-of-health-explore-impact-of-climate-change-on-human-health, accessed September 20, 2013.

USA Today. 2013. Poll: Americans want U.S. to prepare for climate change. March 28. http://www.usatoday.com/story/news/nation/2013/03/28/poll-climate-change/2028223/, accessed July 6, 2013.

US Senate. 2005. Energy Policy Act of 2005 Engrossed Amendment Senate. Section 1612. Subparagraph (b). June 28. http://www.gpo.gov/fdsys/pkg/BILLS-109hr6eas/pdf/BILLS-109hr6eas.pdf, accessed September 22, 2013.

US State Department. 2011. Strengthening U.S.-China sub-national cooperation: The U.S.-China governors forum. Fact sheet. January 19. http://www.state.gov/r/pa/prs/ps/2011/01/154874.htm, accessed July 8, 2013.

Wesoff, Eric. 2012. Adobe adds 400 kilowatts of Bloom fuel cells—but can the fuel cell firm survive as subsidies fade? Greentechmedia, February 2. http://www.greentechmedia.com/articles/read/adobe-adds-400-megawatts-more-of-bloom-fuel-cells, accessed September 1, 2013.

West Coast Infrastructure Exchange. 2012. Framework to establish a West Coast infrastructure exchange (WCX). November 14. http://westcoastx.com/assets/documents/WCX_framework-agreement.pdf, accessed September 20, 2013.

White House. 2001. Letter from President George W. Bush to Senators Hagel, Helms, Craig, and Roberts. Press release, March 13. http://georgewbush-whitehouse.archives.gov/news/releases/2001/03/20010314.html, accessed July 8, 2013.

White House Council on Environmental Quality. 2013a. Council on Environmental Quality press releases. http://www.whitehouse.gov/administration/eop/ceq/press_releases.

White House Council on Environmental Quality. 2013b. Interagency carbon capture and storage task force. http://www.whitehouse.gov/administration/eop/ceq/initiatives/ccs

White House Council on Environmental Quality. 2013c. National ocean policy implementation plan. http://www.whitehouse.gov/administration/eop/ceq/initiatives/oceans.

Wong, Edward. 2013. China's Plan to Curb Air Pollution Sets Limits on Coal Use and Vehicles. September 13. *New York Times*. National Edition, A4.

World Resources Institute and World Business Council for Sustainable Development. 2013. Greenhouse Gas Protocol: City and Community GHG Accounting. http://www.ghgprotocol.org/city-accounting, accessed September 19, 2013.

Xu, Nan, and Chun Zhang. 2013. What the world is getting wrong about China and climate change. China Dialogue. February 18. http://www.chinadialogue.net/article/show/single/en/5711-What-the-world-is-getting-wrong-about-China-and-climate-change, as cited in Andrew Revkin, Tough truths from China on CO_2 and climate, *New York Times* Dot Earth blog, February 26, 2013.

Zhou, Nan, David Fridley, Michael McNeil, Nina Zheng, Jing Ke, and Mark Levine. 2011. China's energy and carbon emissions outlook to 2050. China Energy Group, Lawrence Berkeley National Laboratory. http://china.lbl.gov/sites/china.lbl.gov/files/2050_Summary_Report_042811_FINAL.pdf.

Zillman, John W. 2009. A history of climate activities. WMO Bulletin. 58 (3) July. http://www.wmo.int/pages/publications/bulletin_en/archive/58_3_en/58_3_zillman_en.html, accessed September 19, 2013.

6

Climate Change as News: Challenges in Communicating Environmental Science

Andrew C. Revkin

A few decades ago, anyone with a notepad or camera could have looked almost anywhere and chronicled a vivid trail of environmental despoliation and disregard. Only a few journalists and authors, to their credit, were able to recognize a looming disaster hiding in plain sight.

But at least it was in plain sight. Now, the nature of environmental news is often profoundly different. Biologists these days are more apt to talk about ecosystem integrity than the problems facing eagles or some other individual charismatic species. The subject of sprawl is as diffuse and diverse as the landscapes it encompasses. Concerns about air pollution have migrated—from the choking plumes of old to the smallest of particles that penetrate deep into the lungs and to the invisible heat-trapping greenhouse gases linked to global warming, led by innocuous carbon dioxide, the bubbles in beer. Even though scientists say the main cause of recent warming is smokestack and tailpipe emissions, projections of the pace and ramifications of future climate changes remain as murky as the mix of clouds, particles, and gases that determine how much sunlight reaches the earth and how much heat radiates back into space—the balance that sets the global thermostat.

The challenges encountered in meaningfully translating such issues for the public today are enormous for a host of reasons. Some relate to the subtlety or complexity of the pollution and ecological issues that remain after glaring problems have been addressed. Others relate to effective, well-financed efforts by some industries and groups that oppose pollution restrictions to amplify the uncertainties in environmental science and exploit the tendency of journalists to seek two sides to any issue. This approach can effectively perpetuate confusion, contention, and ultimately public disengagement and inaction.

On the other side of the debate, environmental groups are not innocent in this regard. In some cases, they have focused media attention on their favored issues by going beyond the data and magnifying the risks of, say, cancer or abrupt climate change. Some scientists, expressing frustration with the public's indifference to long-term threats, have stepped outside their areas of expertise and portrayed warming as a real-time catastrophe.

The rhetoric swelled in the spring of 2006 as documentary films, books, and magazine cover stories endeavored to directly link the outbreak of hurricanes, and particularly the ferocity of Hurricane Katrina, to the slow buildup of heat in the world's oceans from human activities. *Time* magazine proclaimed on April 3 "Be worried. Be very worried" (Kluger 2006). A trailer for *An Inconvenient Truth*, the film that documents former Vice President Al Gore's peripatetic multimedia climate campaign, called it "the most terrifying film you will ever see."

Many climate experts said that while there was a growing likelihood that humans were helping shape storm patterns and the like, the inherent variability and complexity of the climate system guaranteed that drawing any straight lines was impossible. On hurricanes, for example, even some of the scientists who claim to have found a relationship between rising hurricane intensity and human-caused warming said that no one could point (with any credibility) to this relationship affecting a particular storm or season.

Critics of those who proclaimed the dawn of a real-time man-made climate catastrophe lashed out. In an opinion piece in the *Wall Street Journal*, Richard S. Lindzen (2006), a climatologist at the Massachusetts Institute of Technology who has long disputed the dominant view that humans could dangerously warm the climate, labeled some calamitous claims "lies" and derided what he called an "alarmist gale."

Between the depictions of global warming as an unfolding catastrophe and as a nonevent lies what appears to be the dominant and still troubling view: that the buildup of carbon dioxide and other long-lived greenhouse gases poses a sufficient risk of profound and largely irreversible transformations of climate and coastlines to warrant prompt action to limit future harms. That view was clearly articulated by eleven national academies of science, including the US National Academy, in a letter to world leaders in 2005. In 2011, the National Academy of Sciences Committee on America's Climate Choices concluded: "the environmental, economic, and humanitarian risks of climate change indicate a pressing need for substantial action to limit the magnitude of climate change and to prepare for adapting to its impacts" (NAS 2011, 1).

Many experts explain that it is urgent to act promptly to curb emissions and limit future risks. In fact, because of population growth and increased energy use in developing countries, even the most optimistic scenarios project that concentrations of greenhouse gases will continue to climb throughout the first half of the twenty-first century.

The problem is that the processes that winnow and shape the news have a hard time handling the global warming issue in an effective way. The media seem either to overplay a sense of imminent calamity or to ignore the issue altogether because it is not black and white or on a timescale that feels like news. This approach leaves society like a ship at anchor, swinging cyclically with the tide and not going anywhere.

What is lost in the swings of media coverage is a century of study and evidence that supports the keystone findings: human-generated gases are trapping heat, and the ongoing buildup of this greenhouse gas blanket adds to warming, shrinks the world's frozen zones, raises seas, and shifts climate patterns.

Certainly, the disinformation generated on both sides of the issue can trip up even earnest, skilled journalists. And the complexity of climate science and policy questions poses a huge challenge in media that are constrained by deadlines and a limited supply of column inches or newscast minutes. Another hurdle is the persistent lack of basic scientific literacy on the part of the public. Nonetheless, some of the biggest impediments to effective climate coverage seem to lie not out in the examined world but back in the newsroom and in the nature of news itself. Overcoming these impediments is a persistent and daunting task. No one should expect to pick up a daily paper anytime soon and read a headline that takes climate science across some threshold of definitiveness that will suddenly trigger public agitation and policy action—and if such a story does appear, it should be looked at skeptically.

A Legacy of Calamity

A little reflection is useful. Most journalists of my generation were raised in an age of imminent calamity. Cold-War "duck-and-cover" exercises regularly sent us to school basements. The prospect of "silent springs" hung in the wind. We grew up in a landscape where environmental problems were easy to identify. The shores of the Hudson River, for example, were coated with adhesives, dyes, and paint, depending on which riverfront factory was nearest, and the entire river was a repository for human waste, making most sections unswimmable. Across the United States, smokestacks were unfiltered. Gasoline was leaded. Los Angeles air was beige.

Then things began to change. New words crept into the popular lexicon—*smog, acid rain, toxic waste*. At the same time, citizens gained a sense of empowerment as popular protests shortened a war. A new target was pollution. Earth Day was something new and vital, not an anachronistic notion. Republican administrations and bipartisan Congresses created laws and agencies aimed at restoring air and water quality and protecting wildlife. And remarkably, those laws began to work.

Still, through the 1980s, the prime environmental issues of the day—and thus in the news—continued to revolve around iconic incidents that were catastrophic in nature. First came Love Canal, quickly followed by Superfund cleanup laws. Then came Bhopal, which generated the first right-to-know laws granting communities information about the chemicals stored and emitted by nearby businesses. Chernobyl illustrated the perils that were only hinted at by Three Mile Island. The grounding of the *Exxon Valdez* powerfully illustrated the ecological risks of extracting and shipping oil in pristine places. Debates about wildlife conservation generally focused on high-profile species like the spotted owl or whales, and gripping stories in which a charismatic creature was a target of developers or insatiable industries presented simplistic views of reality.

In the late 1980s, the world began to focus on the harm caused by burning in the Amazon and other tropical forests. Forest destruction was made personal and relevant to citizens of the industrialized world when the forests were portrayed as the "lungs of the world" or our "medicine chest"—not because scientists suddenly found a way to describe the extraordinary biological diversity of rain forests and the role they play in the global climate system.

Indeed, the first sustained media coverage of global warming was spawned not by a growing recognition that long-lived emissions from industrial smokestacks and tailpipes could alter the climate. Instead, it began when the American public experienced a record hot summer in 1988 just as satellites and the space shuttle were transmitting images of thousands of fires burning across the Amazon basin. The burning season in the rain forests was unleashing torrents of carbon dioxide that were perceived as directly perilous to us, so we paid attention. These days, deforestation in the tropics is once again a distant regional issue and has faded to near-obscurity in the press—resurging only briefly when someone prominent is gunned down there, like the American nun Sister Dorothy Stang in 2005.

Nuclear Winter, Nuclear Autumn

My first stories about the atmosphere and climate came a few years before the scorching greenhouse summer of 1988 and focused on the inverse of global warming—nuclear winter. Here was a ready-made news story. Prominent scientist-communicators, most notably Carl Sagan and Paul Ehrlich, calculated that anyone surviving a nuclear war might perish in the months of cold and dark that followed as the smoke-veiled sky chilled the earth and devastated agriculture and ecosystems. As the scientists met with Pope John Paul II and the theory made the covers of major magazines, the scenario brought new pressure on leaders to find a way to end the Cold War. Within a couple of years, however, fresh scientific analysis showed that the aftermath of nuclear war might be more like a nuclear autumn (to use a phrase coined by Stephen Schneider and Starley Thompson, climate scientists who independently assessed the question). A prediction of nuclear winter was dramatic, dangerous, and novel news. Nuclear autumn was not news, and the double-doomsday scenario quickly faded.

In the meantime, global warming began to build and ebb as a story, always building a bit more with each cycle. If there is one barometer that can help a society gauge whether a problem is real, it is longevity. Unlike concerns about nuclear winter and despite challenges by antiregulatory lobbyists and skeptical scientists, concerns about climate change have not diminished. Instead, evidence of the link and its potential dangers has built relentlessly, as is deftly charted in *The Discovery of Global Warming* by Spencer R. Weart (2008, 2013), a historian at the American Institute of Physics.

An Ozone Hole over Antarctica

In the late 1980s, there was a sense of the new about the greenhouse effect, even though scientists had been positing since the 1890s that heat-trapping gases, particularly carbon dioxide released by burning coal and other focal fuels, could raise global temperatures. A combination of observations and computer simulations seemed finally to be giving a face to theory, which made it easy to sell as a cover story in *Time* magazine or to *Science Digest, Discover*, the *Washington Post*, or the *New York Times*. At that time, there was also a newly perceived global atmospheric threat—the damage to the ozone layer from chlorofluorocarbons and other synthetic compounds—and an international solution in a treaty that banned the chemicals.

But eliminating a handful of chemicals produced by a handful of companies is a very different challenge than eliminating emissions from almost every activity of modern life—from turning on a lamp to driving a car. Another difference between global warming and ozone damage was the iconic nature of the ozone problem. It was an issue with an emblem—the stark, seasonal "hole" that was discovered in the protective atmospheric veil over Antarctica. If a picture is worth a thousand words, a satellite image of a giant purple bruise like a gap in the planet's radiation shield must be worth 10,000. Indeed, according to many surveys, the ozone hole still resonates in the popular imagination—incorrectly—as a cause of global warming simply because it is so memorable and has something to do with the changing atmosphere. The ozone hole also resonated with the public because it was directly linked with an issue that concerns everyone—their health—through the possible risk of increased rates of skin cancer. There, too, global warming is different. Health effects are mostly indirect and involve tradeoffs, such as reduced deaths from severe cold along with increased deaths from heat. Overall, the Intergovernmental Panel on Climate Change in 2007 concluded that, so far, "the effects are small but are projected to progressively increase in all countries and regions" (Confalonieri et al. 2007, 393).

Still, human contributions to the greenhouse effect have remained a perennial issue. Specialized reporters have tracked the developments in climate science and the policy debates over the implications of that science. Tracking scientific progress has become somewhat akin to the old art of Kremlinology—sifting for subtle shifts of language showing that vexing questions are being resolved. Every five years or so, fresh hints emerge from the Intergovernmental Panel on Climate Change (IPCC), the United Nations's scientific body charged with assessing the state of understanding of the problem. The group has sought to be as concrete as possible in its findings, giving quantitative weight to words and phrases such as "likely" and "very likely." That metric has helped the media meaningfully explain the incremental improvements in scientific understanding of the causes and consequences of warming.

The other vital component of the assessment process has been the use of scenarios to depict how certain societal behaviors, particularly energy use, might affect the pace and extent of climate shifts over the course of the century. For the public, this practice provides boundaries for outcomes and a means of judging what kind of response is the most reasonable.

But the incremental nature of climate research and its uncertain scenarios will continue to make the issue of global warming incompatible with the news process. Indeed, global warming remains the antithesis of what is traditionally defined as news. Its intricacies, which often involve overlapping disciplines, confuse scientists, citizens, and reporters—even though its effects will be widespread, both in geography and across time. Journalism craves the concrete, the known, the here and now and is repelled by conditionality, distance, and the future.

If ever there were a moment for a page-one story on climate, for example, it came in October 2000, when a scientist sent me a final draft of the summary for policymakers from the IPCC's third climate assessment, which was due out early in 2001 (Revkin 2000). For the first time, nearly all of the caveats were gone, and there was a firm statement that "most" (meaning more than half) of the warming trend since 1950 was probably due to the human-caused buildup of greenhouse gases. To me, that was a profound turning point, and I wrote my story that way: greenhouse gases produced mainly by the burning of fossil fuels are altering the atmosphere in ways that affect earth's climate, and it is likely that they have "contributed substantially to the observed warming over the last 50 years," as the international panel of climate scientists concluded. The panel said temperatures could go higher than previously predicted if emissions are not curtailed.

This represents a significant shift in tone—from couched to relatively confident—for the panel of hundreds of scientists, the IPCC, which issued two previous assessments of the research into global warming theory, in 1995 and 1990.

To the *New York Times*, this was just another news story, and it was outcompeted for the front page by presidential politics, the breakup of AT&T, the overthrow of an Ivory Coast junta, a study on the value of defibrillators in public places, and a decision by Hillary Clinton to return some campaign contributions from a Muslim group. Reporters, scientists, and the public can take steps to improve this situation. The first one is simply to anticipate the hurdles that can create trouble when the news media and climate science mix.

The Tyranny of News

A fundamental impediment to coverage of today's top environmental issues is the nature of news. News is almost always something that happens that makes the world different today. A war starts. A tsunami strikes. In contrast, most of the big environmental themes of this century concern phenomena that are complicated, diffuse, and poorly understood, with harms spread over time and space. Runoff from parking lots, gas stations, and driveways invisibly puts the equivalent of one and a half *Exxon Valdez* loads of petroleum into coastal ecosystems each year, the National Research Council's Committee on Oil in the Sea (2003) found. But try getting a photo of that or finding a way to make an editor understand its implications. A journalism professor of mine once spoke of the "MEGO" factor: "my eyes glaze over." I've seen that look come over more than a few editors in my years of pitching stories on climate.

Climate change is the poster child of twenty-first-century environmental issues. Many experts say that it will be a defining ecological and socioeconomic problem in a generation or two, and that actions must be taken now to avert a huge increase in emissions linked to warming as economies in developing countries expand. But you will never see a headline in a major paper reading "Global Warming Strikes: Crops Wither, Coasts Flood, Species Vanish." All of those things may happen in plain sight in coming decades, but they will occur so dispersed in time and geography that they will not constitute news as we know it.

Most changes in the landscape and developments in climate science are by nature incremental. Even as science clarifies, it also remains laden with statistical analyses, including broad *error bars*. In the newsrooms I know, the adjective *incremental* in a story is certain death for any front-page prospects, yet it is the defining characteristic of most environmental research. Editors crave certainty: hedging and caveats are red flags that immediately diminish the newsworthiness of a story.

In fact, reporters and editors are sometimes tempted to play up the juiciest—and often least certain—facet of some environmental development, particularly in the late afternoon as everyone in the newsroom sifts for the "front-page thought." They do so at their peril and at the risk of engendering even more cynicism and uncertainty in the minds of readers about the value of the media—especially when one month later the news shifts in a new direction. As a reporter, it can be hard to turn off one's news instinct and insist that a story is not "frontable" or that it deserves 300 words and not 800, but it is possible—kind of like training yourself to reach for an apple when you crave a cookie.

Scientists have gotten into trouble for doing the same kind of thing. Over and over, I meet scientists who despair that issues they see as vital, like climate change or diminishing biological diversity, are not receiving adequate attention. They feel that they "get it" and the rest of the world does not. When talking to the media, some have been tempted to push beyond what the science supports—focusing on the high end of projections of global temperatures in 2100 or highlighting the scarier scenarios for emissions of greenhouse gases. A few scientists and environmental groups linked Florida's devastating 2004 and 2005 hurricane seasons to warming, even though the inherent variability in hurricane frequency and targets precludes any such link without a host of caveats, and scientific projections call only for slight intensification of tropical storms late in the century, not greater numbers.

The coverage linking these storms to warming oceans resulted in a backlash when some hurricane experts disputed the assertions made to the media. Some statements made to the press about climate and hurricanes were made by climatologists who lacked expertise in the conditions generating these great storms. As a result, in 2004 one federal hurricane expert, Christopher Landsea, withdrew in protest from the climate-review process at the IPCC, leading to stories about a dispute over climate science. The result was probably more public confusion and cynicism about what is going on.

This tendency of everyone, from scientists to reporters, to focus on the most provocative element when climate becomes news backfired in a very big way in August 2000. A science reporter for the *New York Times* wrote that a couple of scientists on a tourist icebreaker cruise in the Arctic had seen a large patch of open water at the North Pole, possibly the first such occurrence in thousands of years. Better yet, there were pictures. In an interview, one of the scientists ascribed the open water to global warming, and on a quiet summer weekend, the story popped to the top of the front page (Wilford 2000). Finally, the climate-change issue seemed to be behaving like a news story. It was vivid and dramatic, implying that profound changes were afoot. Television reports and political cartoonists quickly followed up with items on the loss of Santa's summer residence.

Unfortunately, the story was incorrect. Calling a few independent experts might have helped the reporter to avoid trouble. From the fall of 2007 forward, it has been evident that the Arctic is seeing a major reduction in sea ice. However, the sighting in 2000 was unremarkable. Floating sea ice is always a maze of puzzle pieces and open areas. Society would have to wait for its global warming wakeup call.

After covering climate for over twenty years, my sense is that there will be no single new finding that will generate headlines that galvanize public action and political pressure. Even extreme climate anomalies, such as a decade-long super drought in the West, could never be shown to be definitively caused by human-driven warming.

The Tyranny of Balance

Journalism has long relied on the age-old method of finding a yea-sayer and a nay-sayer to frame any issue, from abortion to zoning. It is an easy way for reporters to show they have no bias. But when dealing with a complicated environmental issue, this method is also an easy way to perpetuate confusion in readers' minds about issues and about the media's purpose. When this format is overused, it tends to highlight the opinions of people at the polarized edges of a debate instead of in the much grayer middle, where consensus generally lies. The following maxim illustrates the weakness of this technique: "For every PhD, there is an equal and opposite PhD." The practice also tends to focus attention on a handful of telegenic or quotable people working in the field who are not necessarily the greatest authorities. There are exceptions, but over the years I have learned to be skeptical of scientists who are adept at speaking in sound bites.

One solution to the tyranny of balance is for writers to cultivate scientists in various realms—chemistry, climatology, oceanography—whose expertise and lack of investment in a particular bias are well established. These people can operate as guides more than as sources to quote in a story. Another way to avoid the pitfall of false balance is to focus on research published in peer-reviewed journals rather than that announced in press releases. Peer review, as scientists know all too well, is a highly imperfect process. But it provides an initial quality-control test for new findings that advance understanding of an issue.

The norm of journalistic balance has been exploited by opponents of emissions curbs. Starting in the late 1990s, big companies whose profits were tied to fossil fuels recognized they could use this journalistic practice to amplify the inherent uncertainties in climate projections and thus potentially delay cuts in emissions from burning those fuels. Perhaps the most glaring evidence of this strategy was a long memo written by Joe Walker (1998), who worked in public relations at the American Petroleum Institute, a group that surfaced in 1998. According to this "Global Climate Science Communications

Action Plan," first revealed by my colleague John Cushman at the *New York Times*, "Victory will be achieved when uncertainties in climate science become part of the conventional wisdom" for "average citizens" and "the media" (Cushman 1998). The action plan called for scientists to be recruited and given media training in how to highlight questions about climate while downplaying evidence pointing to dangers. Since then, industry-funded groups have used the media's tradition of quoting people with competing views to convey a state of confusion, even as consensus on warming has grown.

An analysis of twenty years of newspaper coverage of global warming, including articles in the *New York Times*, showed how the norm of journalistic balance actually introduced a bias into coverage of climate change. Researchers from the University of California at Santa Cruz and American University tracked stories that portrayed science as being deadlocked over human-caused warming, being skeptical of it, or agreeing it was occurring. While the shift toward consensus was clearly seen in periodic assessments by the IPCC, the coverage lagged significantly and tended to portray the science as not settled (Boykoff and Boykoff 2004).

One practice that can improve coverage of climate and similar issues is what I call "truth in labeling." Reporters should discern and describe the motivations of the people cited in a story. If a meteorologist is also a senior fellow at the Marshall Institute, an industry-funded think tank that opposes many environmental regulations, then the journalist's responsibility is to know that connection and to mention it. Such a voice can have a place in a story focused on the policy debate, for example, but not in a story where the only questions are about science. The same would go for a biologist working for the World Wildlife Fund.

Another effective approach is to listen carefully to the facts embedded in what someone is saying, regardless of that person's affiliations. For a 2003 story on the politicization of climate science, for example, I interviewed Patrick J. Michaels, a University of Virginia climatologist at the time and outspoken critic of the mainstream view that human-caused warming is dangerous. While laying out his argument against that view, he said he had recently calculated that the most likely warming in the twenty-first century would be just 1.5°C (2.7°F). Later, I realized that Michaels—a prime skeptic who received income through his affiliation with the Cato Institute, an antiregulatory group that was supported substantially by energy companies—had in fact essentially entered the mainstream. His predicted warming was more than two and a half times twentieth-century warming and within the range projected by the IPCC.

None of this comes easily, in part because of two more hurdles that constrain a reporter's ability to characterize what is being said in a story.

The Twin Tyrannies of Time and Space

I came to newspapers after writing magazine stories and books and at first was petrified about filing on a daily deadline. One of my editors, hovering over my shoulder and alluding to the stately pace of other forms of publication, while daylight ebbed, gently put it this way: "Revkin, this ain't no seed catalog." Through the ensuing years, I adapted to the rhythm of the daily deadline but also to the reality of its limitations. On an issue like the environment, I understood why the crutch of "on the one hand" was so popular: there is often simply no time to canvass experts. I grew to understand why stories tend sometimes to read like a cartoon version of the world. There is just no time to do better.

And then there is the question of space. Science is one of the few realms where reporters essentially have to presume the reader has no familiarity at all with the basics, particularly something as complicated as climate science. Just about anyone in America knows the rules of politics, business, baseball, and other subjects in the news. But studies of scientific literacy show that most people know little about atoms, viruses, or the atmosphere. So a lot of extra explication somehow has to fit into the same amount of space devoted to a story on a stock split, a primary vote, or a ball game—and it doesn't. Stories about global warming are not granted a few hundred extra words because it is harder than other subjects.

The shrinking of a climate story that is competing on a page with national or foreign developments is as predictable as the retreat of mountain glaciers in this century. But the material that is cut matters to researchers and to those who want to convey the real state of understanding: the caveats, the couching, the words like *may* and *could*, the new questions that emerge with every answer. Labeling ideally should be there to characterize the various voices in a story.

The only solution is to educate editors as much as possible about the importance of context and precision in such stories. That fight is getting more difficult as the media feel more pressure to generate profits and attract readers. More and more, the limited "news hole" reserved for science in newspapers is being filled with stories on subjects most likely to boost circulation, like fitness, autism, diet, and cancer. That leaves ever-fewer columns for basic science or research on looming risks like climate change.

The Tyranny of Diminishing Resources

Journalism is facing declining resources as more coverage moves to the Internet. This shift has offered new means for telling stories, including video and interactive graphics, but has come with a steep and continuing decline in revenue for newspapers, magazines, and other media.

The result is a steady erosion of the capacity to sustain accurate coverage of issues that are outside conventional subject areas like finance, sports, politics, and security. Climate change may have enormous importance in the grand, century-scale scope of the human enterprise, but it is not top news day to day, month to month. Covering the stock market or the Oscars is a no-brainer. But media resources devoted to coverage of climate change and, to some extent, energy are in decline.

In 2009, the *New York Times* created an environment "pod" distinct from the science desk, with two editors and a team of reporters dedicated to tracking water, energy, biodiversity, climate science and policy, and the like. In 2013, that operation was closed, with some reporters shifted to other beats and those still focused on the environment moved to the science desk or elsewhere. Lost in the shift was the Green blog, which had effectively aggregated and analyzed environmental news. And the *New York Times* is pretty much the best-case scenario compared to other newspapers. The number of reporters in the United States whose primary job is covering climate science can be counted on a single hand.

A number of Internet-based climate science news resources are aiming to fill the gap. Climatecentral.org is one example. Inside Climate News, which won the 2013 Pulitzer Prize for national reporting (for a series on oil-pipeline troubles), is another.

RealClimate.org is an evolving effort by some scientists to convey science directly to the public. This blog is a vital touchstone in the Web conversation on global warming, and represents another instance of reshaping the discourse around climate news and climate science into more of a multidimensional conversation.

The news vacuum is also being filled by online enterprises with a wide range of policy agendas, from ClimateDepot.com, backed by the Center for a Constructive Tomorrow, which is devoted to limited regulation and backed by industry and conservative donors, to Climateprogress.org, maintained by the Center for American Progress, a left-leaning nonprofit organization that seeks tight regulations on greenhouse gases.

Heat versus Light

One of the most difficult challenges in covering the environment, especially when faced with declining resources, is finding the appropriate way to ensure a different kind of balance—between the potent "heat" generated by emotional content and the "light" of science and statistics. Consider a cancer cluster. A reporter constructing a story has various puzzle pieces to connect. Some paragraphs or images brim with the emotional power of the grief of a mother who lost a child to leukemia in a suburb where industrial effluent once tainted the water. A dry section lays out the cold statistical reality of epidemiology, which might never be able to determine if contamination caused the cancer. No matter how one builds such a story, it may be impossible for the reader to come away with anything other than the conviction that contamination killed.

In the climate arena, substitute drowning polar bears or displaced Arctic cultures for cancer-stricken children, and you have the same dynamic at work. It is vital to explore how a warming climate affects ecosystems and people. But this tactic can backfire if a story downplays the uncertainties surrounding unusual climate events or if it portrays everything unusual in the world today as driven by human-caused warming.

It is my impression that the European press, which gives more attention than American media do to climate, has also been more apt to play up hot content and minimize the cooler elements that might deflate a story's sense of drama. This approach caught hold in the United States after Hurricane Katrina (Gore 2006), to the extent that the poster for *An Inconvenient Truth* showed a plume from a smokestack merging with a swirling satellite image of a hurricane. When "superstorm" Sandy flooded Wall Street and coastal regions around New York City, the resulting disaster created a burst of coverage of climate change, both through the influence on such storm surges from rising sea levels and the idea that reduced Arctic sea ice and warming ocean waters affected the storm's evolution and track. While sea level rise is clearly connected to climate change, the other factors remain complicated and uncertain.

The climate connection made for gripping TV and powerful headlines, none more so than in the *Bloomberg Business-Week* article titled: "It's Global Warming, Stupid: If Hurricane Sandy doesn't persuade Americans to get serious about climate change, nothing will." Are media that adopt this approach doing their job? By the metrics that matter in the newsroom, the answer is probably yes. Pushing the limits is a reporter's duty. Finding the one element that's new and implies malfeasance or peril is the key to getting on the front page.

I hope that my own work and that of others will try to refine purely news-driven instincts, to understand and convey the tentative nature of new scientific knowledge, and to retain at least some shades of gray in all that black and white. We also need to drive home that once a core body of understanding has accumulated over decades on an issue—as is the case with the basic aspects of human-driven climate change—society can use it as a foundation for policies and choices.

The Great Divide

Journalists dealing with global warming and similar issues would do well to focus on the points of deep consensus, generate stories containing voices that illuminate instead of confuse, convey the complex without putting readers (or editors) to sleep, and cast science in its role as a signpost pointing toward possible futures, not as a font of crystalline answers.

The only way to accomplish this is for reporters to become more familiar with scientists and the ways of science. This requires using those rare quiet moments between breaking-news days to talk to climate modelers, ecologists, or oceanographers who are not on the spot because their university has just issued a press release. By getting a better feel for the breakthrough and setback rhythms of research, a reporter is less likely to forget that on any particular day the state of knowledge about endocrine disruptors, PCBs, or climate is temporary. Readers will gain the resolve to act in the face of uncertainty once they absorb that some uncertainty is the norm, not a temporary state that will give way to magical clarity sometime in the future.

There is another reason to do this. Just as the public has become cynical about the value of news, many scientists have become cynical and fearful about journalism. Some of this is their own fault. When I was at a meeting in Irvine, California, on building better bridges between science and the public, one researcher stood up to recount her personal "horror story" about how a reporter misrepresented her statements and got everything wrong. I asked her if she had called the reporter or newspaper to fix the errors and begin a dialogue about preventing future ones. She had not even considered doing so.

Cynical unconcern for the presumed failings of journalism in part prolonged the career of the disgraced former *New York Times* reporter Jayson Blair. Few of the people who identified falsehoods in his stories called the paper to correct them. The interactions between sources, journalists, and readers ideally should take on more of the characteristics of a conversation. The communication of news cannot remain effective if it is a monologue.

The more scientists and journalists talk, the more likely it is that the public—through the media—will appreciate what science can (and cannot) offer as society grapples with difficult questions about how to invest scarce resources. An intensified dialogue of this sort is becoming ever more important as science and technology increasingly underpin daily life and the progress of modern civilization.

Given the enormous consequences and irreversible losses from global warming should the worst projections play out, the time for improving the flow of information on this subject is clearly now.

Gatekeepers and Communication

In the past, information and news were analyzed and disseminated by experts and the media for consumption by the public. Increasingly, the preparation and packaging of information is becoming decentralized. And that process itself—whether the preparation of a peer-reviewed journal article or correspondence between experts—is becoming more transparent. These shifting dynamics can create confusion or greater understanding—or both.

In 2009, over a thousand emails between scientists of many nations were uploaded to the Internet without the permission of their authors. These emails revealed the contentious, complex, and often uncertain nature of climate science, and accusations of deception and illegality abounded online and in the mainstream media. "Climategate," as the incident came to be known, provided ammunition for partisans eager to cast doubt on arguments for action to curb greenhouse gases, even while polls have shown few lasting changes in public concern about climate change. The incident appeared to reinforce distrust of scientists among some people in the United States, "primarily among individuals with a strongly individualistic worldview or politically conservative ideology" (Leiserowitz et al. 2010, 1). A suite of inquiries cleared the scientists of any wrongdoing, and a police investigation in Britain ended without finding out who released the email caches.

The core observational datasets that underpin analysis of climate change trends have been reopened and reanalyzed. Demands have proliferated for greater transparency around scientific data, processes, and analytical models. In response, efforts to open-source all climate computer code have advanced. Significant changes in the way the IPCC assesses research and delineates areas of scientific agreement and disagreement have been recommended. In an effort to exhibit transparency and reestablish trust in the scientific process, new guidelines were suggested by the Royal Society. The new guidelines suggest splitting climate science results into three categories distinct from one another: "aspects of wide agreement," "aspects of continuing debate," and "aspects not well understood" (Hulme 2010).

As Mike Hulme, professor of climate change in the School of Environmental Sciences at the University of East Anglia, notes:

There has been a re-framing of climate change. The simple linear frame of "here's the consensus science, now let's make climate policy" has lost out to the more ambiguous frame: "What combination of contested political values, diverse human ideals and emergent scientific evidence can drive climate policy?" The events of the past year have finally buried the notion that scientific predictions about future climate change can be certain or precise enough to force global policy-making. ... The meta-framing of climate change has therefore moved from being bi-polar—that either the scientific evidence is strong enough for action or else it is too weak for action—to being multi-polar—that narratives of climate change mobilise widely differing values which can't be homogenised through appeals to science. It is clearer today that the battle lines around climate change have to be drawn using the language of politics, values and ethics rather than the one-dimensional language of scientific consensus or lack thereof. (Hulme 2010, 2–3)

I know a number of supervising authors of the IPCC reports are eager to revise policies and stress openness. I also understand those who shy away from the media, particularly given how some media overplayed claims that the climate panel had erred in parts of its 2007 assessment.

But any instinct to pull back after being burned by the news process is mistaken, to my mind. As I explained to a roomful of researchers at the National Academy of Sciences, in a world of expanding communication options and shrinking specialized media, scientists and their institutions need to help foster clear and open communication more than ever. Clampdowns on press access almost always backfire.

That's not to say that an open approach to communication with the public will be easy. Such efforts have to account for the tendency of cultural groups with divergent belief systems— think liberals and libertarians, along one axis—to react very differently to the same body of scientific information if it is presented in ways that challenge one group's worldview. In particular, studies by Dan Kahan of Yale University and colleagues suggest that simply providing more scientific information to a polarized public can deepen polarization. They find that a common understanding of scientific information is difficult to achieve if individuals feel the science threatens their ability to remain welcome members of their cultural group. In a *Nature* commentary, Kahan explains, "When all citizens simultaneously follow this individually rational strategy of belief formation, their collective well-being will certainly suffer. Culturally polarized democracies are less likely to adopt polices that reflect the best available scientific evidence on matters—such as climate change—that profoundly affect their common interests" (Kahan 2012, 255).

Of course, there is no simple solution here. The prescription offered by Kahan and others in a 2012 *Nature* Climate Change paper has the feel of a Sisyphean task: "communicators should endeavor to create a deliberative climate in which accepting the best available science does not threaten any group's values" (Kahan et al. 2012).

But this work surely points to the need for experimentation and for using variegated voices and approaches—a task that Randy Olson, a marine biologist who's become a filmmaker and author of books on science communication, calls going "beyond the nerd loop" (Revkin 2011).

The global climate-change science community is still in the early days of refining its approach to outreach and interactivity with the public. But challenges remain for the media and the public as well. There are no gatekeepers anymore. Consumers of online news and analysis have to develop ways of independently assessing the quality of information and assertions.

The Readers' Responsibility

Most people use the Web to search for something already in mind, or rely on aggregators like Drudgereport.com or Treehugger.com to feed them prefiltered news. For this reason, people don't generally come across novel information or perspectives very often. Even when a friend on Facebook or a Twitter contact shares a link, most Facebook friends are almost assuredly likeminded in some ways.

This is very different than twentieth-century pathways for news and information. News reports, whether on a TV screen or front page, had the feel of an orderly public square, wherein different categories of news were conveyed side by side, offering a bazaar-style menu of ideas. For example, a reader might come to the *New York Times* for information on stocks, but would flip through pages and see an article on Greenland's glaciers, as well.

This raises a societal issue: how to help readers maintain a broad public awareness of global events. The Web poses a great paradox. We have access to nearly everything, but we use it to sort for things we focus on already. This means that going forward, to a large extent, consumers of news and information who care about maintaining breadth and quality in the perspectives before them need to take an active role in reaching out for varied sources of information. I encourage my blogging students to follow some people on Twitter whose views differ from their own. I encourage people eager to learn about some development in climate or energy science to try switching from a standard Google search, which ends up tailored to their own biases over time and picks up a blend of opinion and fact, to Google Scholar (scholar.google.com), which sifts academic literature—much of it peer reviewed. As with all online communication, whether putting things out or taking them in, progress will only come with conscious attention to the potential for the ultimate connectivity tool to further isolate and polarize sectors of society.

A New Generation of Journalists

There has been an erosion of a powerful model: a network of correspondents dispensing knowledge from on high. It has been replaced by an interactive mode of journalism. For example, in 2013, I have 110,000 followers on Facebook. That is my biggest audience. The challenge for news consumers today isn't a lack of information—but too much of it. Today, communicators who succeed will be those who can serve as reliable guides in a complex landscape full of tricky terrain. As I've written, the job of a journalist today is more like being a trusted mountain guide after an avalanche. That's very different than the Cronkite-style approach, which ended with "That's the way it is."

To succeed, people starting out in journalism today need to carve out a niche and take ownership of it. Brian Stelter and Nate Silver are valuable examples. They were independent bloggers. They had novel approaches. They had passion. They were hired by the *New York Times*. Now they are stars. They started as independent, talented reporters with a fresh take on an existing question. The skill set for success as a journalist in the world of online reporting can include the ability to translate static data into dynamic visualizations or a video—or the capacity to collaborate with other communicators to do so.

And in conveying the causes and consequences of climate change, and possible personal and societal responses, the time is ripe for fresh experiments. Early on, global warming was depicted as a conventional environmental crisis—which implicitly implied there was some single solution (like a filter on a smokestack). It's clear now that the evolving human response to a changing climate, as noted by Matthew Nisbet of American University (2011, 3; 2013, 4), is more akin to persistent issues like public health or poverty—an area full of persistent and variegated challenges that can only be addressed through sustained focus and dialogue, not some simple set of steps. The outcome, later this century, is likely to resemble that in the Thornton Wilder play "By the Skin of our Teeth." An informed, engaged public is vital if we are to limit regrets in a world facing a changing climate and retreating coastlines for centuries to come, even as some countries continue their climb out of poverty. My guess is we'll pull a Thornton Wilder and get by, even thrive—but the *IF* in the preceding sentence should really be in capital letters and italics.

References

Boykoff, J. M., and M. T. Boykoff. 2004. Balance as bias: Global warming and the U.S. prestige press. *Global Environmental Change* 14:125–136.

Confalonieri, U., B. Menne, R. Akhtar, K. L. Ebi, M. Hauengue, R.S. Kovats, B. Revich, and A. Woodward. 2007. Human health. In *Climate Change 2007: Impacts, Adaptation and Vulnerability. Contribution of Working Group II to the Fourth Assessment Report of the Intergovernmental Panel on Climate Change*, edited by M. L. Parry, O. F. Canziani, J. P. Palutikof, P. J. van der Linden, and C. E. Hanson. Cambridge, UK: Cambridge University Press, 391-431.

Cushman, John H., Jr. 1998. Industrial group battles climate treaty. *New York Times*. April 26, A1.

Gore, Al. 2006. *An Inconvenient Truth: The Planetary Emergency of Global Warming and What We Can Do about It*. Emmaus, PA: Rodale Press.

Hulme, Mike. 2010. The year climate science was redefined. theguardian.com. Posted November 16. http://www.theguardian.com/environment/2010/nov/15/year-climate-science-was-redefined, accessed September 19, 2013.

Kahan, Dan. 2012. Why we are poles apart on climate change. *Nature* 488 (7411):255. doi:10.1038/488255a.

Kahan, Dan M., Ellen Peters, Maggie Wittlin, Paul Slovic, Lisa Larrimore Ouellette, Donald Braman, and Gregory Mandel. 2012. The polarizing impact of science literacy and numeracy on perceived climate change risks. *Nature Climate Change* 2 (10):732–735. doi:10.1038/nclimate1547.

Kluger, Jeffrey. 2006. Be worried. Be very worried. Earth at the tipping point. *Time*. April 3, 24–54.

Leiserowitz, Anthony, Edward W. Maibach, Connie Roser-Renouf, Nicholas Smith, and Erica Dawson. 2010. Climategate, public opinion, and the loss of trust. Working Paper Series. Social Science Research Network. http://ssrn.com/abstract=1633932, accessed September 21, 2013.

Lindzen, Richard. 2006. Climate of fear. *Wall Street Journal*. April 12, A14.

National Academy of Sciences. 2011. *America's Climate Choices*. Committee on America's Climate Choices; National Research Council. Washington, DC: National Academies Press.

National Research Council Committee on Oil in the Sea. 2003. *Oil in the Sea III: Inputs, Fates, and Effects*. Washington, DC: National Academies Press.

Nisbet, Matthew. 2011. *Climate Shift: Clear Vision for the Next Decade of Public Debate*. Recent Studies. Climate Shift Project. American University School of Communication. http://climateshiftproject.org/, accessed September 21, 2013.

Nisbet, Matthew. 2013. *Nature's Prophet: Bill McKibben as Journalist, Public Intellectual and Activist*. Joan Shorenstein Center on the Press, Politics and Public Policy. Discussion Paper Series #D-78. March. John F. Kennedy School of Government, Harvard University. http://shorensteincenter.org/wp-content/uploads/2013/03/D-78-Nisbet1.pdf, accessed September 21, 2013.

Revkin, Andrew C. 2000. A shift in stance on global warming theory. *New York Times*. October 26, A22.

Revkin, Andrew C. 2011. Climate, communication and the "nerd loop." *New York Times,* Dot Earth blog, April 14. http://dotearth .blogs.nytimes.com/2011/04/14/climate-communication-and-the-nerd-loop/, accessed July 8, 2013.

Walker, Joe. 1998. Global Climate Science Communications Action Plan. Memo, American Petroleum Industry. April 3. http://www .climatesciencewatch.org/file-uploads/API_communication_plan_ memo.pdf, accessed July 8, 2013.

Weart, Spencer R. 2008. *The Discovery of Global Warming.* Revised and Expanded Edition. Cambridge, MA: Harvard University Press.

Weart, Spencer R. 2013. *The Discovery of Global Warming: A hypertext history of how scientists came to (partly) understand what people are doing to cause climate change.* February 2013 version. Center for History of Physics of the American Institute of Physics. http://www .aip.org/history/climate/index.htm, accessed September 21, 2013.

Wilford, John Noble. 2000. Ages-old icecap at North Pole is now liquid, scientists find. *New York Times.* August 19, A1.

7

Climate Change and Human Security

Richard A. Matthew

Introduction

In this chapter I argue that in much of the world today, the prospects for human security are linked to the trajectory of climate change. Unfortunately, rather than bringing the world together around a robust global action plan, this growing linkage may well divide the world into two increasingly disconnected solitudes.

I begin by discussing the failures of the United Nations Framework Convention on Climate Change (UNFCCC) process and the fear, especially pronounced within the science community, that we are losing ground and face a very turbulent and alarming future. I then briefly present the concept of human security. A third section explores some of the links between human security and climate change. This sets up an argument that these links may actually divide rather than bring together humankind, a trend I argue is evident in the UNFCCC process itself. I conclude with some suggestions for a different approach to addressing the challenge of climate change.

The Failure of the UNFCCC Process

The UNFCCC was negotiated at Rio in 1992 with the explicit purpose of reducing greenhouse gas emissions in order to reduce human impact on the world's climate system. Over more than twenty years, the UNFCCC has made no progress in fulfilling its purpose, instead devolving into a complex process comprised of many moving parts that, from one year to the next, are as likely to move sideways or backward as they are to move forward. If the planet experiences the climate-driven catastrophes that many scientists foresee, then the image of thousands of negotiators, experts, and lobbyists gathering unsuccessfully late each year to devise and fine-tune strategies for mitigating and adapting to the global challenge of climate change may imprint history as the meme for a particular type of human folly: when great capacity fails to meet great needs because of even greater human selfishness. Since 1995, parties to the agreement, 195 states in 2013, have met in December to address a lengthening list of outstanding issues. These meetings have been bleak and unproductive periods of political theater, with unconvincing villains like Canada and hapless heroes like the world's small-island states, and all parties liberated from the need to make arguments based on evidence or reason.

At the center of current global climate-change negotiations stands the Kyoto Protocol. Negotiated in 1997, it came into effect in 2005 and its first commitment period expired in 2012. Its modest objective of reducing key greenhouse gas emissions by 5 percent below 1990 levels was not reached or even approached; instead, these emissions increased by 58 percent during this period (Paris 2012). The reasons for this failure disclose what are widely perceived as the major fault lines hampering global efforts to address climate change.

First, Kyoto did not require any commitments from the developing world, even though countries like China and India were and are the site of enormous and growing emissions. Second, key wealthy players—the United States and Australia—quickly abandoned the Kyoto process, in part because of the preceding point. And third, countries were allowed, at least informally, to shift the baseline—Canada, for example, has declared progress by using 2005 as its baseline instead of 1990.

Also embedded in the Kyoto Protocol are more general problems that haunt the entire UNFCCC process. In particular, no rigorous methodology has been agreed upon for measuring, reporting, and verifying decreases in emissions. Further, Kyoto is not legitimized through or integrated into any broader vision of how to address climate change—mainly because no such vision exists at the global level. Consequently, much activity at the eighteenth conference of parties (COP 18), on the eve of the Kyoto Protocol's expiration, focused on simply extending this remarkably weak protocol for another eight years. While this extension was approved in the final moments of the meeting, the fact remains that Kyoto does not address any of the bigger issues at play—and, writing in the wake of COP 18, it seems quite possible that the extension itself will soon flounder because it left partly unresolved the serious issue of whether Russia could carry over the large account of emissions allowances it accumulated during the first phase of the protocol due to its sputtering post–Cold War economy.

These bigger issues, of course, were discussed in great detail at COP 18, as has been the case in earlier COPs. The so-called Durban platform, the key output of COP 17, has been assigned the role of bringing the developing world into the UNFCCC process by 2015. It is unclear what this will entail, however, as this integration faces several enormous obstacles. The first has been a thorn in the side of the UNFCCC process since the beginning—how to make actionable the concept of "common but differentiated responsibilities and respective capacities" (abbreviated as CBDR+RC). While at a high level of abstraction this concept makes some sense, it has proven very hard to operationalize. Medium powers like Norway and Germany often argue for strong commitments by rich countries, while big powers like the United States and China are far more restrained in this regard. Until and unless the world can agree on a new interpretation of CBDR+RC that lays the foundation for meaningful action, the UNFCCC process will continue spinning its wheels—and probably causing more global warming than it mitigates.

A second challenge has to do with the controversial issue of measurement, reporting, and verification (MRV). Fortunately, at COP 18, a process for resolving this was set in motion; it will be interesting to see its outputs in the years ahead. This group will certainly have many meetings in exotic locales; but it is doubtful that real transparency and science-based evidence are in the interest of many of the parties to the treaty.

Third, and perhaps most daunting, is the perennial question of financing. While the developed countries have pledged $100 billion to support mitigation and adaptation efforts beginning in 2020, there is little funding secured for the next several years, a potentially critical period of massive planetary growth in urban space, land modification, and energy demand. At COP 18, developing countries deepened their pitch for some sort of "loss and damage" fund, through which they would be compensated for suffering adverse climate-change consequences. It is hard to imagine how climate-change effects would be monetized, however, given the now-standard practice of not attributing any particular event to climate change per se. It is equally hard to imagine the developed world putting much capital into such a fund.

Finally, there is not a shred of evidence that large developing countries like India and Pakistan have any intention of ever making commitments to this regime. The threshold for admission to the UNFCCC has been shamefully low, and developing nations could sign the treaty for PR purposes without much fear that they would ever have to change their behavior.

The one area in which the UNFCCC process has been effective is in identifying new needs, such as the need for flexible National Adaptation Plans, and the need to set up new bureaucracies, such as the Adaptation Committee, to prepare reports on new needs—two of the concrete outcomes of COP 18.

Why has the UNFCCC process been so lackluster, given that the people who participate in it often depict climate change as an existential threat to humankind (Welzer 2012)? Indeed, over the past two decades, the dominant climate-science narrative has resolved into a very concise and alarming story. For some 5,000 years, this story goes, humankind has lived in a period of remarkable climate stability. It has taken advantage of this congenial period to craft vast, complex societies. But these societies require huge amounts of energy to function and grow. The best supply of energy—fossil fuels—was created over millions of years by natural forces, and humankind has relied on this cheap and abundant source to fuel its development. Unfortunately, fossil fuel emissions accumulate in the atmosphere and cause global warming. Climate scientists worry that global warming will hasten us out of the congenial climate period we have known, and into a world of catastrophic droughts, floods, heat waves, and storms. These often will overwhelm the intricate human systems that have been constructed around the provision of food, energy, transportation, public health, and so on, because, at best, they were designed to withstand the lesser fluctuations of weather characteristic of the past five millennia (Pearce 2007; Welzer 2012). In short, much of the built environment may not survive the more aggressive climate that humankind itself is causing, ironically forcing people to move, inflaming and introducing violent conflicts, and widely expanding the domain of human misery and insecurity.

The gap between the terrible vision of a climate-changed world and the anemic outputs of two decades of climate-change negotiations is wide and growing. Often this gap is explained by arguing that the divisions noted above—especially between developed and developing economies and between big and medium powers—holds the process hostage to questions of who is responsible and who should pay for transformation, and thus ensures that outcomes will always be modest. The concept of human security offers another perspective on this question. Human security shifts our focus from the level of humankind and the state, where much of the climate-change negotiation takes place, to the level of the individual and community, where much of the impact of climate change is being experienced. At this scale the incentives to mitigate and otherwise address climate effects can be far more immediate, and hence more likely to catalyze behavioral change and policy innovation. But at this scale the sense that we are all in this together, and need to act in concert, can erode because effects vary enormously from place to place and over time. In other words, human security provides a useful framework for assessing current climate-change effects and encourages strong linkages to local needs and capacities, but can also make very clear how differently climate change is playing out across the surface of the planet.

Human Security

The concept of human security was introduced into global discourse through the 1994 United Nations Development Programme (UNDP) *Human Development Report*. Human security "can be said to have two main aspects. It means, first, safety from such chronic threats as hunger, disease and repression. And second, it means protection from sudden and hurtful disruptions in the patterns of daily life" (UNDP 1994, 23). The UNDP identified seven areas of global concern: economics, food, health, environment, personal security from physical violence, community, and political security (recognition of basic human rights). Climate change has the potential to jeopardize all of these.

In the 1990s, the UNDP report inspired two rather different policy approaches. The first, identified with Canada, focuses on "freedom from fear"; that is, on protecting people from explicit forms of violence ranging from landmines through human trafficking to civil war. Its strategies include humanitarian assistance, peace-building, and conflict prevention and mediation, and its signature campaigns have focused on things like banning the use of landmines, outlawing child soldiers, and shining a spotlight on human trafficking. The second approach, identified with Japan, focuses on "freedom from want," and hence on the far more elaborate mandate of working to ensure all people have an opportunity to enjoy lives of security, welfare, and dignity. Strategies here are largely focused around development assistance, and signature campaigns include pursuing the Millennium Development Goals.

Perhaps because it has evolved along these two trajectories, the concept of human security has been criticized as too broad to be analytically useful, and, at least in the United States, it certainly has not proven to have the immediate inside-the-belt-way appeal of Robert Kaplan's (1994) "coming anarchy" thesis, which entered academic and policy discourse at precisely the same time. Nonetheless, its development has been steady, and it has demonstrated considerable attraction from scholars, policymakers, and activists in the developing world and Europe.[1]

Tariq Banuri, for example, offers a concise argument in defense of human security:

Security denotes conditions which make people feel secure against want, deprivation, and violence; or the absence of conditions that produce insecurity, namely the threat of deprivation or violence. This brings two additional elements to the conventional connotation (referred to here as political security), namely human security and environmental security. (1996, 163–64)

In this conception, structural insecurities and violence associated with the world economy and the legacies of colonialism, together with modalities of violence and insecurity associated with environmental change—two force fields that are themselves interactive and historically related—combine to ensure that large portions of humankind, primarily in the South but not exclusively so, are rarely, if ever, free from danger and want.

The fact that "human security" embodies a great deal may make it less analytically interesting to some, but it would be wrong to suggest that there is not much analytical value in broad inclusive concepts that tell a compelling general story.[2]

In his analysis of the concept, Roland Paris notes that such a high level of inclusiveness can "hobble the concept of human security as a useful tool of analysis," but he ultimately concludes that:

Definitional expansiveness and ambiguity are powerful attributes of human security ... human security could provide a handy label for a broad category of research ... that may also help to establish this brand of research as a central component of the security studies field. (2001, 102)

Vulnerability is often used as a way of assessing human security and giving it greater analytical traction. Vulnerability can mean simply exposure to a hazard, but typically it involves some measure of adaptive capacity and other systemic properties. Vulnerability is thus commonly regarded as a function of exposure, sensitivity, adaptive capacity, and mitigation capacity, and thus might be rendered as $Vf(E \times S \times A \times M)$. Much of this particular effort to focus the concept of human security and use it as a basis for analysis and policy making has been undertaken by scholars and activists in the field of environmental security, a field that recently has focused considerable attention on climate change. (e.g., Adger 1999; Floyd and Matthew 2012; Khagram et al. 2003; Lonergan 1999; Matthew et al. 2009; Nauman 1996).

Human Security and Climate Change

It is perhaps important for the integrity of the UNFCCC process to foster a sense of shared vulnerability, and hence of shared fate, through the idea that climate change is a global challenge with the potential to negatively affect virtually everyone on the planet. But in fact, with the evidence at hand, we can imagine three quite different futures.

The first receives much of the attention these days. In this scenario, the negative impacts of climate change increase dramatically throughout the twenty-first century, affecting more and more people, more and more often, and more and more harshly. Severe weather events like Hurricanes Katrina and Sandy, which caused hundreds of deaths and billions of dollars in damage, become acknowledged as manifestations of climate change as they ravage coastal communities around the world. Heat waves continue to plague cities from Paris to Karachi. Droughts turn sub-Saharan Africa and large areas of the United States and China into dust bowls. And, perhaps, global events like the melting of the Greenland ice sheet and the sudden release of vast stocks of methane gas suddenly disable the ocean conveyer that warms much of Europe and disrupt the monsoon upon which the livelihoods of billions of people in Asia depend. All of humankind is cast into a turbulent and inhospitable climate, and people everywhere are hard-pressed to survive. Humankind is finally and tragically connected across geographic, cultural, and economic divides.

In a second scenario, however, the social effects of climate change would simply prove to be far less than expected. Given the vast domain of uncertainty that clouds this area of research, scientists have been remarkably shrill in verbalizing terribly bleak forecasts. But unanticipated phenomena such as increased cloud cover might blunt global warming over the next few decades and establish new equilibria that maintain a fairly stable and congenial global climate regime. Or, the benefits of climate change might offset many of the costs as growing seasons and access to natural resources expand in large areas of the world. And, where climate change effects are truly pernicious, human ingenuity might devise quite effective coping and adaptation responses.

The problem with this scenario, of course, is that it would be wildly imprudent to use it as the basis for policy design and behavioral change. If it is correct, and we respond aggressively to the potential threat of climate change, then our worst outcome is that we will have overspent on new infrastructure, energy efficiency, food production, and water management. But if it is wrong, and we have nonetheless adopted it as a guide for our behavior, then our worst outcome is that we enter the first scenario with no preparation at all.

Finally, in a third scenario, the negative effects of global warming continue to fall disproportionately onto the world's poor and most vulnerable communities. Over time, two solitudes emerge and consolidate. In one, defined by acute vulnerability to climate change, a hardscrabble existence becomes even worse as floods and droughts assume unmanageable proportions, unleashing unprecedented famines and pandemics. Angry and miserable people turn to violence, but often discover that wealth continues to have the upper hand in this regard. They are steadily confined to their miserable domains, isolated from hope, condemned to a Hobbesian existence—short, nasty, brutish. In the second solitude, however, the vulnerability is far less, the climate-change effects are far more mixed, enormous wealth buys time, and innovation ramps up. The focus in this part of the planet is on risk management, on devising entrepreneurial solutions that make money while doing good, on improving efficiency and stimulating innovation, on decoupling from costly foreign problems, and, ultimately, on uncompromising perimeter defense.

No one can predict the future, but I believe the evidence we have to date tends to provide the greatest support for scenario three. Two examples might illustrate this.

Climate-Induced Flooding in Bangladesh

The conventional climate change scenario for Bangladesh focuses on the impact of sea level rise on a coastal region made extremely vulnerable because of the twin pressures of a reduction in the siltation that constitutes a natural barrier to flooding, due in part to water (and hence silt) being diverted upstream into dams and irrigation, and enormous population pressures. Bangladesh has "one of the most densely populated, low-lying, coastal zones in the world, with 20–25 million people living within a one-metre elevation from the high tide level" (Energy and Resources Institute 2005).

This vulnerability enables enormous human security problems to manifest in very short periods. For example, during the monsoon season of 2004, the coastal region of Bangladesh was paralyzed by severe flooding. By mid-July, 60 percent of the country was under a blanket of water soiled with a rank mixture of industrial, agricultural, and household waste. Some 20,000,000 people were directly affected, many of them facing food and fresh water shortages, skin infections, disease, and displacement.

Flooding on a similar scale had occurred sixteen years earlier, but the cost of building a system of dams and dykes that would protect the country was deemed prohibitive by the government, so little was achieved to safeguard the population or its critical infrastructure. Combating climate change, the likely cause of the heavy precipitation, snow melt, and sea level rise that contributed to the flooding, was equally out of reach. Complex processes of global environmental change that amplify severe weather events and widespread poverty have combined to ensure that the people of Bangladesh remain highly vulnerable to all the adverse effects of severe flooding (World Bank 2000).[3]

Figure 7.1 illustrates the potential impact of sea-level rise on Bangladesh.

At the rate of sea level rise that was expected at the time the Bangladesh analysis was conducted (1989), it was expected that the illustrated inundation levels would occur in roughly 150 years, or around 2140. However, sea levels around the

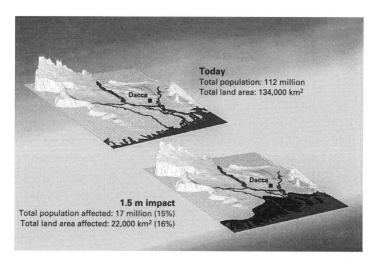

Figure 7.1
Potential impact of sea-level rise in Bangladesh.
Source: United Nations Environmental Programme Global Resource Information Database (Geneva); University of Dacca; JRO (Munich); World Bank; World Resources Institute (Washington DC).

globe have been rising significantly faster than previously anticipated. As a result, it is likely that this level of inundation will occur earlier than previously believed.

But the vulnerability of the country is manifest each year in many different and less dramatic ways. For example, Tanguar Haor is a 9,727-hectare area of rich biodiversity that is protected as an internationally recognized wetland, also known as a Ramsar site after its treaty name, but that also serves as the home to 25,000 of the poorest people in the country. Their average per capita income is a mere $130 US—about one third of the national average in Bangladesh, and about 2 percent of the global average. Organized into forty-six villages that date back to at least the eighteenth century, the people are completely dependent on wetland resources—fish, reeds, forest products—in order to survive. The *haor* itself is formed of numerous ponds known as *beels,* which are inundated during the wet season when they merge into a single large lake. During this period, most of the inhabitants of the area migrate out, returning in the dry season to fish the *beels,* gather grass to create a variety of goods for themselves and for the market, and collect stones to sell for construction.

In the past sixty years this area has changed its national status three times—from Britain to Pakistan to Bangladesh—but throughout this, de facto control has remained in the hands of the region's local elites. Most of the people living in the area have depended on informal, customary agreements for access to the resources of the *haors.* These agreements are far less respected today than they were in the past. In recent years, the combined effects of colonial structures of inequality, rapid environmental change, dynamic population growth, a leasing system implemented in the 1970s that most local people could not afford and that hence eroded customary rights, and the state's decision to protect the area's biodiversity have all rendered the poor extremely vulnerable to a host of risks, and created an atmosphere that is prone to human insecurity, violence, and misery. As people struggle to survive, deprived of customary access to wetland resources because of government regulation, population growth, and environmental degradation, the area has steadily become volatile and prone to violent conflict.[4]

Increasingly fierce competition over access to the resources of the *haor* is subtly affected by climate change, insofar as it amplifies the wet and dry seasons, making *beels* more or less resource-rich and hence contributing to the creation of incentives that undermine traditional livelihoods, insurance systems, and dispute-resolution mechanisms. All too often the economic and political transformations that occur—partly in response to environmental change—distribute costs to the poor and benefits to the rich. Facing abundance, for example, the rich find ways to channel resources and other goods into the market to make money, at the expense of the livelihoods of the poor. Faced with scarcity, the rich find ways to monopolize what is available, also at the expense of the poor.

Climate-Induced Drought in Sudan

Sudan, independent since 1956, has a population of 39 million and is, like Bangladesh, one of the poorest countries in the world. While Bangladesh has spent a considerable amount of its GDP on maintaining a large military because of the fear of aggression from India, Sudan has spent heavily on a twenty-one-year civil war that ended only in December 2004. But as the civil war came to a close, regional violence in the northern state of Darfur escalated. The economist Jeffrey Sachs writes:

Failures of rainfall contribute not only to famines and chronic hunger, but also to the onset of violence when hungry people clash over scarce food and water. When violence erupts in water-starved regions such as Darfur, Sudan, political leaders tend to view the problems in narrow political terms. If they act at all, they mobilize peacekeepers, international sanctions and humanitarian aid. But Darfur, like Tigre, needs a development strategy to fight hunger and drought even more than it needs peacekeepers. Soldiers cannot keep peace among desperately hungry people. (2005, 1)

Northern Darfur is one of three states in western Sudan. It has a population of about 1.5 million people, who subsist primarily through farming and herding. Population growth has dramatically increased population density in the region, placing enormous pressure on its arid lands. In recent years, declining rainfall, likely a manifestation of global warming, has added considerably to the region's woes. Historically, the region was neglected by the more populous and oil-rich south, a neglect due partly to ethnic prejudice and partly to the civil war that has plagued the country for much of its existence. Ironically, as the civil war has diminished in recent years, culminating in the December 2004 peace agreement, years of drought appear to have intensified long-standing conflicts over access to land and water between the farmers, who are also predominantly black and Christian, and the herders, who tend to identify themselves as Arab Muslims.[5]

As the conflict matured into a government-supported genocide perpetrated by Arab *janjaweed* (armed horsemen) against black farmers, who have antagonized the government by forming two dissident groups (the Sudan Liberation Army and the Justice and Equality Movement), survival strategies have included selling land, produce, and livestock cheaply so that these things will not be stolen, and working the land unsustainably to extract nutritional value—the region suffers from a severe food shortage—as quickly as possible. As the market serving the region has collapsed, poverty and malnutrition have intensified, and local residents have become highly vulnerable to diseases including malaria, yellow fever, cholera, and diarrhea. Since 2003, hundreds of thousands of people have died or been displaced.

In *Hegemony or Survival*, Noam Chomsky writes: "Humans have demonstrated ... [destructive] capacity throughout their history, dramatically in the past few hundred years, with an assault on the environment that sustains life, on the diversity of more complex organisms, and with cold and calculated savagery, on each other as well" (2003, 2). Sudan is a particularly harsh and unsettling case in point.

As in many parts of the world, conflict in Sudan is linked strongly to economic and ethnic factors operating at the local level. But corrupt government, ethnic rivalry, intransigent poverty, and an arid environment also create great vulnerability to the transnational problem of climate change, which enters into Sudan as drought and very quickly amplifies both the vulnerability of the people and the multiple forms of violent conflict and abject misery to which they are more or less permanently exposed. In Sudan, as in Bangladesh, the sustaining elements of vulnerability need to be addressed, so that the people can adapt to or act to mitigate any adverse effects of climate change.

Statistically minded scholars may argue that many more cases are needed before a pattern can be identified that is not pure speculation, but let us speculate here that there is a pattern emerging, one that is at once historically familiar and deeply unsettling.

Two Solitudes

The familiar pattern that informs the two cases is one in which the costs of change are displaced onto the poor and weak, and the benefits of change are seized, often violently, by the rich and the powerful—whose unsustainable practices and values usually provided the rationale for change in the first place. Although developed countries and elites around the world have been stung by storms, droughts, and floods, the scale of suffering is quite different from that of the world's least-advantaged peoples in places such as the Sudan, Pakistan, Bangladesh, or Ethiopia. Ironically, the frontline countries today include small-island states, which came together as a lobby group for the COP meetings but have been unable to agree on what they should be demanding of the world. Some type of compensation, yes, but on what legal basis and for what purpose? Do they want to be moved? Or do they want technological interventions that will allow them to stay on for as long as possible? As they bumble along at COP meetings, bigger powers have begun to think about what the implications might be for the vast expanses of ocean that disappearing small-island states control through the Law of the Sea Convention. Could these enormous exclusive economic zones be leased or sold? Or will they default to coastal states? Tensions are already mounting in this arena among countries such as China, Japan, and South Korea.

In a wonderfully fresh approach to the great elements of diversity that undergird the shared experience of global climate change, Harald Welzer explores "the impact of climate change on global inequalities and living conditions" and concludes that "whether wars in the twenty-first century are directly or indirectly due to climate change, violence has a great future ahead of it" (2012, 6). If climate change makes it more and more difficult for some groups to survive, let alone flourish, then Welzer suggests that no approach to solving this problem will be off the table—including the use of force, by both sides. "It is a modernist superstition that allows us to keep shrinking from the idea that, when people see others as a problem, they also think that killing them is a possible solution" (Welzer 2012, 23). Internationalized civil wars, terrorism, genocide, and the often brutal scramble to control natural resources in Africa demonstrate the continued willingness of both state and nonstate actors to resort to the most brutal forms of violence. But, Welzer argues, the academic world tends to underestimate the potential for violence. This is in part because "violence is only to a very limited degree, if at all, part of the experiential world of the academics who concern themselves with it. As a result, little explicit research has been devoted to this central area of human action, and even that is overloaded with moralism and fantasy" (89).

From Welzer's perspective, climate change already is placing enormous pressure on some people—mainly on those who are poor and vulnerable, as demonstrated in the cases of Bangladesh and the Sudan. Climate science predicts much worse to come. Indeed, the journalist Fred Pearce captures a widely shared sentiment in the science world when he writes that climate change "scares me, just as it scares many of the scientists I have talked to—sober scientists, with careers and reputations to defend, but also with hopes for their own futures and those of their children, and fears that we are the last generation to live with any kind of climatic stability" (2007, xxvi). Given the slow pace of international climate-change negotiations, there is ample reason to speculate that the gap between the planet's needs and its institutional responses is growing, which may have the effect of increasing the likelihood of violence.

In this context, several reasons can be suggested to explain why so little is being done at the international level. The first has to do with climate science itself. It is funded primarily by developed states that want to know what is happening and what is in store for them. Although the rapid accumulation of climate science has led to a broad consensus that global warming is real and caused by human activity—hence the scared scientist phenomenon noted by Pearce—it has failed to catalyze public pressure or political will.

This might be in part because of the character of public discourse:

Climate scientist: Our models predict an average of 0.2°Celsius of global warming per decade to the end of the century.

Developed country citizen: That does not seem like very much.

Climate scientist: Well, that's an average. In some places it will be much higher, maybe 0.4°.

Developed country citizen: Where?

Climate scientist: Mainly where it is already hot—parts of Africa, the Middle East, and South Asia.

Developed country citizen: And what will this increase do?

Climate scientist: It could cause more heat waves, droughts, severe storms, flooding, and over time, as ice sheets melt, sea level rise and disruptions of the ocean conveyer that would have planetary impacts.

Developed country citizen: What does that mean for us?

Climate scientist: Possibly more famines, mass migrations, pandemic disease.

Developed country citizen: Where?

Climate scientist: Certainly in parts of Africa, the Middle East, and South Asia. Potentially anywhere.

Developed country citizen: Are you certain about this?

Climate scientist: Pretty certain about global warming caused by human actions. Less certain about how this will play out in weather over the next few decades, and how this will affect humans in different regions.

Developed country citizen: I see.

One problem with this type of discourse is that it does not create a space for meaningful action at the individual level—which can only matter if it is aggregated on a huge scale. Thus it firmly places addressing the problem into the realm of international negotiations. But twenty years of little progress in this arena may be signaling to individuals and markets that the problem cannot possibly be that pressing. And so, a certain complacency has settled in, and little pressure is placed on international elites by citizens and corporations to do more than they have done. Another problem is that the claims of science are not being seen as authoritative. The Intergovernmental Panel on Climate Change (2007, 2) has stated that, as of 2007, the planet had warmed by $0.74°C$ ($1.3°F$) in the course of a century, and would continue to warm by about $0.2°C$ ($0.4°F$) per decade until about 2027, at which point the effects of whether mitigation efforts were serious or not would become relevant. The 2013 assessment, as is reported elsewhere in this volume, concluded that "global surface temperature change for the end of the twenty-first century is *likely* to exceed 1.5 degrees C relative to 1850 to 1900" under a number of sets of assumptions and "more *likely than not* to exceed" $2°C$ ($3.6°F$) another scenario (emphasis in original; IPCC 2013, SPM-12). But these claims have been challenged by some skeptics in the United States.

Underlying all of this may be the sense that the urgency revealed by climate science is mainly located in unfortunate parts of the developing world, where change is unlikely because so many things are going badly.

Which leads to a second factor. Without much corporate or social pressure, it is hard to imagine that international elites will ever do more than they are doing now. The UNFCCC process has become a huge and impressive bureaucratic process involving thousands of people, but bureaucracies are designed to administer policy, not to think outside the box, experiment with solutions, and work toward fundamental social transformations. It is unlikely that much pressure will come from within the process for another reason as well. The experts in this process have, for the most part, no experience at all with the adverse impacts of climate change. In fact, they happily jet set around the world, drive SUVs, own large houses, and work in air-conditioned offices and labs.

Although a few token farmers and pastoralists are always on hand to tell their sad stories, negotiations are led by people with little or no personal stake in actually solving the problem of climate change. To the contrary, they operate almost exclusively in a matrix that rewards designing quick-impact policies, protecting particular interests, and agenda setting. And the most powerful voices in the process represent those places where the effects of climate change are mixed or muted, the publics have other priorities, and strong positions might create serious liability issues down the road. In other words, climate change has developed into an issue that can only be addressed through the strong commitments of those who have little or no incentive to make strong commitments, and who feel no pressure at all to change this position.

Uncertainty is a third factor mitigating against progress through the UNFCCC process. A logical response to uncertainty is risk management. Managing climate risk is unlikely to pull people together; uneven impacts swamp common efforts under CBDR+RC (regarding responsibilities and capacities). What the United States or Canada might choose to do to manage their climate risks over the next few decades might tighten and expand their bilateral relationship, but it is hard to see how this could be linked in other than rhetorical ways to the needs of poor communities in India or Ethiopia.

As the implications of climate change emerge, it becomes ever more imperative that we base political decisions related to climate change on environmental justice. If we fail to do so, the impacts of climate change will continue to be unevenly borne by the already vulnerable communities that have historically been most affected by environmental pollution and by climate-related events such as flooding and drought.

Climate Change and Environmental Justice in the Developing World

Even in the most prosperous of developed nations, already vulnerable communities and people living in poverty often will be harmed first and worst by climate change. In developing nations and on the international level, the consequences of climate change will manifest as land lost as a result of sea level rise, crop loss, and famine due to extreme heat and drought in agricultural lands; an increased incidence of disease due to the spread of disease-carrying insects and other vectors; and extreme weather conditions including heat waves, droughts, very heavy rains, and floods, to cite only a few. These impacts will vary from region to region, among income groups and occupations, and by gender. In many cases, the effects of climate change will result in the migration of huge numbers of vulnerable people.

Little recognition or concern for the protection of "climate refugees," those fleeing the impacts of climate change in their native regions, has thus far been expressed by the international community. Land loss, drought, and persistent flooding will result in unknown numbers of people who will be forced to relocate to other regions or countries in an attempt to ensure their survival. While the UN's Guiding Principles on Internal Displacement, international human rights legislation, and various national laws protect displacees within their own countries, those who are forced to relocate across national borders have historically had no such protections. While a new global initiative aimed at providing humanitarian options for international displacees was initiated in 2012 by the countries of Norway and Switzerland (the Nansen Initiative), other nations and government agencies are reluctant to begin yet another international process in an attempt to address this problem.

The Nansen Initiative continues to attempt to build consensus among nations on how to help people displaced from their homelands by sudden climatic shocks such as extensive flooding due to extreme storm events, or by climate impacts that develop more slowly, such as prolonged drought or sea level rise (IRIN 2013a).

Within the developing world, certain population groups are likely to suffer more from the effects of climate change. Throughout the world, women form a disproportionately large sector of the population currently living in poverty. In the developing world, women depend heavily on local natural resources for their livelihoods and are the ones responsible for providing water, food, and cooking and heating fuel for their families. In Africa, 80 percent of women make their living through agriculture. Drought, deforestation, and changes in rainfall patterns and totals will adversely impact their ability to provide these essentials. In addition, women in the developing world have little access to political forums and therefore have little power to influence the decisions of policy makers (Women Watch n.d.). The Council for Development of Social Science Research in Africa held a conference on Gender and Climate Change in November 2012, and determined that "the feminization of poverty and the dominance of patriarchal values in Africa" means that African women will be inordinately impacted by climate change (Darcel 2013). In 1979, the UN General Assembly adopted the Convention on the Elimination of All Forms of Discrimination against Women (CEDAW). CEDAW comprehensively defines discrimination against women and sets forth an action agenda for eliminating discrimination against women. More than one hundred nations have ratified the convention; nations that have consented to or ratified the convention are legally required to abide by its terms. The convention provides a basis for achieving gender equality through insuring equal access and opportunities for both men and women in public and political life, and in the fields of health, education, and employment (United Nations Division for the Advancement of Women 2009). In nations that abide by the terms of the convention, women should have a greater voice in the creation and implementation of climate-change policy.

Children also are, and will continue to be, disproportionately impacted by climate change and natural disasters. Studies in Ethiopia, Kenya, Niger, and India demonstrate that children who are born during droughts or floods are more likely than other children to be malnourished. The effects of climate change will very likely worsen drought and flood conditions in these nations and in other developing nations that are already fighting to combat hunger and food insecurity. A conference held in Dublin, Ireland, in 2013 looked at the issue of climate change from a "climate justice" perspective, focusing on ensuring the rights of vulnerable populations who are least responsible for, and most impacted by, climate change. The conference suggested adopting a "nexus approach" to address interconnected climate change and malnutrition problems: this type of approach requires cooperation and coordination across health, environment, agriculture, and water and land management agencies (IRIN 2013b).

Climate Change and Environmental Justice in the United States

In the United States, the most vulnerable communities are indigenous peoples, people of color, and the poor. These population groups are already at a disadvantage socioeconomically, are burdened by poor environmental quality to a far greater extent than are more prosperous communities, and are least able to reduce or adapt to current and future environmental challenges or health risks. For example, the American Lung Association reported in 2009 that 80 percent of Latinos and 71 percent of African Americans lived in areas that did not meet air quality standards (Shepard and Corbin-Mark 2009). One method commonly proposed in the United States to reduce greenhouse gas (GHG) emissions and improve air quality in general is the cap-and-trade approach.

Environmental Justice and Cap-and-Trade Programs

Cap-and-trade programs may offer a cost-effective mechanism to reduce GHG emissions. In such a program, the government establishes a cap, or a limit on the total amount of GHGs that a covered industry can emit. The government then creates emission allowances equal to the established limit (cap) and distributes them to the firms covered by the program. Allowances may be freely given or auctioned to the highest bidder. Covered firms are then permitted to trade allowances with other covered firms. One advantage of this system is that when a covered firm determines that cutting emissions would be prohibitively expensive, it can buy allowances to continue, rather than reduce, its total emissions. It does this by paying another firm to make the required reductions in its place. Using this system, the collective costs of achieving specified emissions limits can be reduced (Kaswan 2009).

While the overall benefits of reducing GHGs would be equitably distributed under such a system, the costs of cap-and-trade programs may fall disproportionately on communities surrounding the production facilities unless the program incorporates safeguards to prevent this outcome. GHG-generating facilities such as coal-fired power plants or petroleum refineries also generate other pollutants, such as sulfur dioxide, volatile organic compounds, particulate matter, benzene, and other toxic or harmful substances. If facilities within vulnerable communities are able to continue their production of GHGs by purchasing allowances from other firms, these additional pollutants, or "co-pollutants," would continue to affect the surrounding neighborhoods. In the future, as the nation turns to other forms of energy production, emissions of GHGs and their associated pollutants will diminish in all communities; however, in the short term, unless environmental justice policies reduce co-pollutant emissions, the persistent issue of nonattainment of national air quality standards and related public health impacts in certain areas may fail to be addressed. In addition, if allowances can be purchased on an international basis from developing nations, the emissions of GHG—and the generation of co-pollutants—could continue unabated within developed nations. One solution to this potential problem would be to require facilities within the United States and other developed nations to reduce emissions of GHG and co-pollutants instead of permitting the purchase of domestic or international allowances or offsets (Kaswan 2009).

Based on environmental justice concerns, and out of concern for the potential failure of cap-and-trade programs to adequately address the problem of climate change, many within the environmental justice community have scorned cap-and-trade programs as a means of reducing GHG emissions and addressing climate change. Others, acknowledging that cap-and-trade programs will most likely be incorporated into national climate-change mitigation strategies—and may even form the cornerstone for US policy aimed at mitigating climate change—are proposing methods to maximize the potential benefits and minimize the potential risks of cap-and-trade programs (Kaswan 2009).

A cap-and-trade program is a vital component of California's Global Warming Solutions Act (AB32 2006). The program covers major GHG emission sources including refineries, power plants, industrial facilities, and transportation fuels. The California Air Resources Board (ARB) will distribute tradable allowances equal to the overall emissions permitted under the cap, which will decrease over time. Well-designed GHG reduction programs, however, do incorporate other strategies in combination with cap-and-trade. The approach used in California is to combine a cap-and-trade program with more traditional regulatory strategies to reduce GHG emissions, including direct regulations, alternative compliance mechanisms, monetary and nonmonetary incentives, voluntary actions such as reductions of GHGs that occur prior to being mandated under the law, and a program implementation regulation to finance the program. The ARB published a draft update in 2013. The update will assess progress on goals of the 2008 AB32 Scoping Plan and analyze how California's longer-term GHG emission reduction strategies may be aligned with other key policy designs for natural resources, water use, clean energy and transportation, land use, and waste management (California ARB 2013).

The ARB considers ensuring environmental justice, which California defines as "the fair treatment of people of all races, cultures, and incomes with respect to the development, adoption, implementation, and enforcement of environmental laws, regulations, and policies" an integral part of its endeavors. To that end, the ARB has adopted policies to ensure the legally mandated incorporation of environmental justice into its programs. In addition, AB32 instructed the ARB to create a global warming environmental justice advisory committee to advise the board in any matter pertinent to implementing AB32. The ten-member advisory committee includes representatives from California communities where exposures to air pollution are most severe. These communities include, among others, communities with minority or low-income populations. Committee meetings are open to the public and include a period dedicated to hearing the comments of the public (California ARB 2013).

The ARB's policies cover the full spectrum of all ARB actions and apply to all communities in California; however, they also reflect the understanding that environmental justice issues tend to be raised to a greater extent in low-income and minority communities. Underpinning the ARB approach is the recognition that community members must be engaged in a meaningful way in the design and implementation of ARB strategies and activities. In order to participate meaningfully, people need access to the best available information about the air in their communities and about what steps are or may be taken to improve air quality. In order to implement its policies, ARB works closely with communities, environmental and public health organizations, business owners, industry representatives, other state and local agencies, and all other interested parties (California ARB 2013).

Climate Change and Environmental or Social Justice Organizations

Throughout the United States, numerous national, regional, and local environmental and social justice organizations are working to influence government policies related to climate change. One of those organizations is the Environmental Justice Climate Change Initiative (EJCC), which was founded in

2001. Since that time, the EJCC has become a voice for justice and equity in domestic and international climate-change discussions. For example, the EJCC has partnered with EPA and others to hold public workshops on environmental justice (EJCC 2012).

Environmental justice organizations often form networks to broaden their reach and promote proposals in regional and local policy. The Environmental Justice Leadership Forum on Climate Change defines itself as "a national coalition of environmental justice organizations working together to advance climate justice and impact policy to ensure the protection and promotion of communities of color and low-income communities throughout the U.S." (2009).

In a letter to the Bicameral Task Force on Climate Change dated February 26, 2013, one of the forum's member organizations, WE ACT for Environmental Justice, presented the Forum's policy recommendations in response to the task force's request for their responses related to the following climate-change policy issues: actions or policies federal agencies could take to reduce emissions of heat-trapping pollution (mitigation); actions or policies federal agencies could adopt to make the United States more resilient to the effects of climate change (adaptation); and recommendations for congressional legislation aimed at strengthening the ability of federal agencies to prevent and respond to the effects of climate change. Under the category "Federal Agency Actions or Policies," for example, the Forum offered fifteen action recommendations for regulations, permitted activities, or funding programs. Their recommendations were wide-ranging and included the following suggestions: establishing regulations that encompass the notion of co-pollutants and hotspots; requiring health impact assessments as part of NEPA requirements; developing funding programs for "smart grid" infrastructure improvements and for the development of microgrids that use and store clean energy; and requiring "a fee on carbon emissions that will encourage emission sources to invest in more efficient controls while ensuring that carbon emissions are being reduced." Under the category "Administrative policies," their ten recommendations included such items as engaging environmental justice communities in discussions as advocates for vulnerable populations; setting aside funding and resources to address the mental health challenges of climate refugees; and ensuring that "equity is a central component of clean energy and green infrastructure development to ensure that the benefits of these programs are significantly extended to low income and/ or communities of color." Under the category "Congressional legislation," an additional ten recommendations included the creation of legislation that would "hold federal agencies accountable for their climate adaptation plans," commissioning Government Accounting Office reports on the negative effects of climate change on vulnerable population groups, and creating legislation that forbids the purchase of carbon allowances by polluters in or near vulnerable communities (Environmental Justice Leadership Forum on Climate Change 2013).

Climate Change and Human Security: What Does the Future Hold?

Of course, perhaps there will be a breakthrough in the UN-FCCC process, and perhaps the Durban platform will succeed in truly amazing ways. But if this does not happen, if the future looks a lot like the past, then we can imagine that climate-change effects will continue to aggressively undermine human security in several regions of the world—predominantly in poor, fragile, post-colonial regions like Sub-Saharan Africa, the Middle East, and South Asia. As water evaporates, food systems collapse, disease spreads, and people move, Welzer's fears could materialize.

My vision of this outcome is that two solitudes already visible may become more concrete, imperfect geographies connected by a few real and virtual bridges but largely experiencing the impacts of climate change on their own. The rich parts of the world will cooperate, innovate, adapt, and overcome. Their investors who are exposed in the areas where the impacts are most severe—companies that have vast mining concessions and governments that have purchased or leased huge tracts of land for food or biofuel production—will face violence and will have to contemplate defending their interests through violence. If they are smart, they will have reduced their own risks by inviting other investors into their portfolios, building leverage for using force if other strategies fail.

As for the desperately poor, they may well find that the walls that keep them out of the more stable and prosperous parts of the world have grown alongside global warming, often invisibly thanks to powerful surveillance and communication technologies.

The emergence of two solitudes would not mean that the first scenario, in which the effects of climate change eventually overwhelm us all, could not also prove true. My concern is that two solitudes may consolidate in the years ahead, making the first scenario more likely.

Conclusions

Welzer concludes *Climate Wars* with an appeal for "*a critique of any limitation of survival conditions for others*" (2012,

163, emphasis in original). It is possible that self-interest alone might support this critique—no one wants to be the victim of a disease or act of terrorism or crime that has its origins in some desperate and impoverished community whose land has been rendered uninhabitable by forces of environmental destruction that are beyond its control. Or the critique might be informed by a moral imperative to promote human security globally, rooted in human rights discourse or some other universalizing moral argument such as Fritjof Capra's (1997) notion that we are united through the "web of life."

The critique is emerging in several places. For example, Mike Brklacich, the former director of the Global Environmental Change and Human Security Program, argues that it is unclear that the Framework Convention on Climate Change and the Kyoto Protocol will succeed in halting or even slowing human-generated climate change. But even if we cannot stop trends like global warming, at least in the near future, we can and should act to reduce the vulnerabilities of the world's least advantaged people, for it is through these vulnerabilities that climate change becomes a significant threat to safety and welfare. But the declaratory tone of this and many similar statements, while uplifting, does not make clear the motivation for taking action.

Today, approximately one-fifth of humankind survives on 1 percent of the world's product, while another fifth lives on some 80 percent of it. In spite of enormous and virtually universal gains in areas such as infant and child mortality, life expectancy, and literacy—all variables used to measure human security and human development—equally enormous asymmetries of power and wealth have taken shape or been reinforced in the last few decades. The combined wealth of the richest ten people on the planet is greater than that of the poorest 20 percent of the world's population. We may be more connected to each other than ever before, but our circumstances vary dramatically, and we are not equally vulnerable to transnational forces such as climate change.

In this context it is easy for the rich to focus on narrowly defined threats and leave solving the far more elusive issue of vulnerability to future generations. The plight of the poor in Bangladesh or in Sudan, of people driven from their homes and their livelihoods by floods or drought that have been amplified by climate change, may be unsettling to watch on CNN, but the television set can be turned off and the images displaced by more immediate—and attractive—ones. Nonetheless, for decades, the growing interconnectedness of the world has compelled us to ask: how vulnerable might we be to climate change or to the spill-over impact of its adverse social effects? Where will these desperately poor people, unemployed and landless and also locked into bloody class or ethnic struggles, turn for help? And what moral obligation, if any, might we have when a process of global change in which we are deeply implicated places great burdens on people who probably have had a fairly negligible impact on the global change itself?

I worry that this sense of interconnectedness, of shared fate, might be vulnerable to a very different vision, one that divides the world into climate-change entrepreneurs and climate-change victims, and that allows two solitudes to develop along increasingly independent trajectories.

Notes

1. See, for example, Lonergan 1999; Thomas and Wilkins 1999; Tehranian 1999; Suhrke 1999; Paris 2001; Yuen 2001; Khagram, Clark, and Raad 2003; and Matthew 2012. A more explicit union of environmental security and human security is evident in Nauman 1996.

Kaplan's thesis is that the type of anarchy he believed was evident in West Africa in the 1990s was due to a combination of demographic change, environmental degradation, the dismantling of national borders, and technological innovations affecting the character of warfare—a set of interactive variables that, over time, will tend to overwhelm governance institutions worldwide.

2. Concepts such as "class relations," "human rights," and "democracy" are broad and inclusive and do an enormous amount of work in contemporary political analysis.

3. This is in stark contrast to low-lying areas in other regions, such as the Netherlands, which Bangladesh turned to for assistance in the 1990s. Unfortunately, it was not able to afford the infrastructure the Dutch have successfully constructed to protect their land. Similarly, the record rainfall in arid Southern California in 2004–2005 caused little harm because over $200 million had been invested in flood control infrastructure following the heavy rains of 1997–98.

4. On ecological collapse, see Diamond 1994.

5. For an important and influential quantitative study linking environmental change to other factors to better explain and predict violent conflict, see Collier 2000.

References

Adger, Neil. 1999. Social vulnerability to climatic change and extremes in coastal Viet Nam. *World Development* 27:249–269.

Banuri, Tariq. 1996. Human security. In *Rethinking Security, Rethinking Development*, edited by Naqvi Nauman, 163–164. Islamabad: Sustainable Development Policy Institute.

California Air Resources Board. 2013. AB 32 Scoping Plan. http://www.arb.ca.gov/cc/scopingplan/scopingplan.htm

Capra, Fritjof. 1997. *The Web of Life*. New York: Anchor Books.

Chomsky, Noam. 2003. *Hegemony or Survival*. New York: Henry Holt.

Collier, Paul. 2000. Economic causes of civil conflict and their implications for policy. A paper for the World Bank Group. http://www-wds.worldbank.org/external/default/WDSContentServer/WDSP/IB/2

004/03/10/000265513_20040310161100/Rendered/PDF/28134.pdf, accessed September 22, 2013.

Darcel, Denise. 2013. African women particularly impacted by climate change. Must be involved in forming strategies, says UN. *Epoch Times.* April 23. http://www.theepochtimes.com/n3/21345-african-women-particularly-impacted-by-climate-change, accessed July 7, 2013.

Diamond, Jared. 1994. Ecological collapse of past civilizations. *Proceedings of the American Philosophical Society* 138:363–370.

Energy and Resources Institute. 2005. Climate Change and Development: South Asia Report. Report No. 2004GW35. In *Climate Change and Development Consultation on Key Researchable Issues*, edited by Saleemul Huq and Hannah Reid. Section 4: South Asia Region. Section 4.1: South Asian Regional Scoping Study, TERI. London: Climate Change Group, International Institute for Environment and Development. http://pubs.iied.org/pdfs/G00055.pdf, accessed July 8, 2013.

Environmental Justice and Climate Change Initiative. 2012. The EJCC partners with The Environmental Protection Agency to host Intergenerational Workshop on Environmental Justice. Posted on July 27. http://www.ejcc.org/theejcc-partners-with-the-environmental-protection-agency-to-host-intergenerational-workshop-on-environmental-justice/, accessed September 22, 2013.

Environmental Justice Leadership Forum on Climate Change. 2009. http://weact.org/Portals/7/Full%20member%20list%20June%20 2009.pdf, accessed July 7, 2013.

Environmental Justice Leadership Forum on Climate Change. 2013. Letter to the Bicameral Task Force on Climate Change. February 26. http://weact.org/LinkClick.aspx?fileticket=b2KhvUf8LBU%3d&tab id=331, accessed July 7, 2013.

Floyd, Rita, and Richard Matthew, eds. 2012. *Environmental Security: Frameworks for Analysis.* Oxford: Routledge.

Intergovernmental Panel on Climate Change. 2007. *Climate Change 2007: Synthesis Report. Summary for Policymakers.* http://www.ipcc .ch/pdf/assessment-report/ar4/syr/ar4_syr_spm.pdf, accessed July 7, 2013.

Intergovernmental Panel on Climate Change. 2013. Working Group I Contribution to the IPCC Fifth Assessment Report Climate Change 2013: The Physical Science Basis Summary for Policymakers. http://

www.climatechange2013.org/images/uploads/WGIAR5-SPM_Approved27Sep2013.pdf.

IRIN. 2013a. Lifeline to "climate refugees"? April 17. http://www.irinnews.org/report/97862/lifeline-to-climate-refugees, accessed July 7, 2013.

IRIN. 2013b. A unified approach to climate change and hunger. April 24. http://www.irinnews.org/Report/97913/A-unified-approach-to-climate-change-and-hunger, accessed July 7, 2013.

Kaplan, Robert D. 1994. The coming anarchy: How scarcity, crime, overpopulation, tribalism, and disease are rapidly destroying the social fabric of our planet. *Atlantic* February. 273 (2): 44–76. http://www.theatlantic.com/magazine/archive/1994/02/the-coming-anarchy/304670, accessed July 7, 2013.

Kaswan, Alice. 2009. CPR Perspective: Environmental justice and climate change. Center for Progressive Reform. http://www.progressivereform.org/perspEJandCC.cfm, accessed July 7, 2013.

Khagram, Sanjeev, William C. Clark, and Dana Firas Raad. 2003. From the environment and human security to sustainable security and development. *Journal of Human Development* 4 (2): 289–313.

Lonergan, Steve. 1999. *Global Environmental Change and Human Security Science Plan. IHDP Report 11*. Bonn: IHDP.

Matthew, R. 2012. Environmental change, human security and regional governance: The case of the Hindu Kush-Himalaya Region. *Global Environmental Politics* 12 (3):100–118.

Matthew, Richard, Jon Barnett, Bryan McDonald, and Karen O'Brien, eds. 2009. *Global Environmental Change and Human Security*. Cambridge, MA: MIT Press.

Nauman, Naqvi, ed. 1996. *Rethinking Security, Rethinking Development*. Islamabad: Sustainable Development Policy Institute.

Paris, Max. 2012. Kyoto climate treaty sputters to a sorry end. CBC News, December 31. http://www.cbc.ca/news/politics/story/2012/12/20/pol-kyoto-protocol-part-one-ends.html, accessed July 7, 2013.

Paris, Roland. 2001. Human security: Paradigm shift or hot air? *International Security* 26 (2):87–102.

Pearce, Fred. 2007. *With Speed and Violence: Why Scientists Fear Tipping Points in Climate Change*. Boston: Beacon Press.

Sachs, Jeffrey. 2005. Climate change and war. http://www.tompaine.com/print/climate_change_and_war.php, accessed July 7, 2013.

Shepard, Peggy M., and Cecil Corbin-Mark. 2009. Climate justice. *Environmental Justice* 2 (4):163–166.

Suhrke, Astri. 1999. Human security and the interests of states. *Security Dialogue* 30 (3):265–276.

Tehranian, Majid, ed. 1999. *Worlds Apart: Human Security and Global Governance*. London: I.B. Tauris.

Thomas, Caroline, and Peter Wilkins, eds. 1999. *Globalization, Human Security and the African Experience*. Boulder, CO: Lynne Reinner.

United Nations Development Programme. 1994. *Human Development Report 1994*. Oxford: Oxford University Press.

United Nations Division for the Advancement of Women. 2009. Convention on the elimination of all forms of discrimination against women. http://www.un.org/womenwatch/daw/cedaw/cedaw.htm, accessed July 7, 2013.

Welzer, Harald. 2012. *Climate Wars: Why People Will Be Killed in the 21st Century*. Cambridge, UK: Polity Press.

WomenWatch. n.d. The threats of climate change are not gender-neutral. http://www.un.org/womenwatch/feature/climate_change, accessed July 7, 2013.

World Bank. 2000. Bangladesh: Climate change and sustainable development. Report No. 21104 BD. Dhaka: World Bank Rural Development Unit.

Yuen, Foong Khong. 2001. Human security: A shotgun approach to alleviating human misery? *Global Governance* 7 (3):231–236.

8

Climate Change: What It Means for Us, Our Children, and Our Grandchildren

Joseph F. C. DiMento, Pamela Doughman, and Suzanne Levesque

Climate change is a complex challenge, perhaps one of the largest the world has ever faced. This book has presented some important findings about climate-change science and related social and policy issues.

We have described changes in climate, including those caused by human activities, and explained why there is a gap between scientific understanding of climate and actions taken by society to remedy the harmful effects of emissions-related climate changes. We have described why climate change has been difficult to understand, based on the characteristics of science as well as communication difficulties and political motivations.

In this book, we described the dynamics of climate and the greenhouse effect, presenting a primer on the science and a historical overview of scientific discoveries. We distinguished between climate and weather and described the components—from clouds to oceans—and interactions that compose the climate system and human contributions to greenhouse gases. We noted *how* we know the climate is changing and *what* humans can do to affect the climate. The earth has become warmer and likely will grow warmer still without major changes in the energy we use and how we use it.

As noted by Hulme (2010), a new pragmatism, based on three principles, now influences the formation of climate policy: a contemporary emphasis on policy changes, such as those related to health or poverty, that also result in benefits to the climate; a need to propel innovative, publicly funded investments in the creation of low-carbon energy technologies; and the formation of broad-based institutions and partnerships for policy innovation at many levels as adjuncts to the UN process.

Climate change poses challenges for local, national, and international governments and the private sector. Further changes in average temperature, precipitation, and weather events will affect human health and global and regional economies. Ecosystems will change, and some species will be made extinct. Many ice sheets will melt or shrink, many glaciers will recede or disappear, and oceans will continue to rise.

Climate change will be different across regions. Agriculture will be more productive in some zones and jeopardized in others. Some areas, like the Arctic, will continue to face dramatic alterations that threaten lifestyles and the viability of flora and fauna. In the United States, changes will be felt in different ways in the Pacific Northwest, California, the Midwest, New York and other parts of the East, and along the many thousands of miles of coast. Changes in weather, water availability, crop yields, heat waves, and public health—and the associated demands for energy—will be significant in some places at even small temperature increases. At the high level of 5.8°C (10.5°F), the impacts will be severe.

In chapter 2, we offered a historical perspective on what scientists know. The impressive research record comes from the work of individual laboratories and the compilations and assessments of existing knowledge by the Intergovernmental Panel on Climate Change (IPCC) and other major scientific bodies. Even so, there continues to be a mistaken view among some of the public that 1) there is no consensus on whether climate change exists; and 2) some scientists have self-interested reasons for viewing the situation with alarm. As discussed in chapter 3, the future of climate change cannot be known with certainty, but it is risky to ignore trends.

Climate scientists—like scientific experts in other disciplines—do not always have expertise when it comes to communicating their findings to the public. Nonetheless, as seen in chapter 4, no matter what some people believe and contrarians assert, the reality of climate change and its effects stands up well to the generally accepted standards of scientific inquiry. Denying that global warming is real is simply a refusal to look at the evidence.

Many individuals, private organizations, states, and national and international groups consider the record sufficiently clear to act, even though some think that more research and analysis are needed before major shifts are made. The latter are driven in part by their assessments of the costs associated with actions to control emissions. In chapter 5, the current and future actions of international efforts such as the Framework Convention on Climate Change and the Kyoto Protocol and "post Kyoto" were examined.

Some countries have gone beyond the minimum requirements of the Kyoto Protocol and have demanded energy efficiency, renewable energy, new products for business and the consumer, and markets for the trading of emissions credits. Others have exited the process, either explicitly or by ignoring the need for controls until more research is done. Some governments have used the legal system (legislation and lawsuits) to force other governments and businesses to decrease their emissions of greenhouse gases. Business has been involved, sometimes by choice, in recognizing the potential sources of profits in emissions-reducing products and processes. The long periods of time that greenhouse gases remain in the atmosphere mean that some level of climate change is already unavoidable, but harmful effects can be mitigated by taking actions—ranging from sustainable forest management and local land use controls to energy-saving residential and commercial building practices.

Communicating information about science is the mission of the Newkirk Center for Science and Society, which sponsored publication of this book. All of the chapters address the difficulties in translating climate-change science to nonscientists and the attentive public, but chapter 6 makes these challenges explicit by comparing them to other topics—including other environmental topics—that clamor for public attention. Some of the major impediments to effective climate coverage lie in the newsroom and in the nature of news itself. Climate scientists are not describing something that has front-page impact. Journalists—who are driven by time, space, and format pressures—at times can confuse readers into believing that human contributions to changes in climate are a matter of great scientific debate. They are not.

The localized effects of climate change will vary. Not everyone will be hurt, and some will likely benefit. But as chapter 7 notes, some regions are extremely vulnerable to ongoing changes in climate, and their populations, their resources, and their environments will be damaged in significant ways. Effects in the truly poorest communities may well be felt worldwide through the economic networks that have evolved in a globalized world.

Building on the insights from chapter 1 through chapter 7, this chapter highlights considerations that influence the meaning of climate change across generations, including the impacts of uncertainty and time on assessments of risk and cost. The chapter concludes by placing the issue in context and providing a final word on actions with no regrets.

Changing Knowledge

Our understandings of the challenges, risks, and opportunities of climate change evolve daily as new scientific information becomes available for society to consider. Abrupt climate change has moved from science fiction to a subject of serious attention. In 2005, one analysis, for example, concluded we have perhaps ten to thirty years to keep carbon levels from doubling and avoid rapid, extreme climate change (Baer and Athanasiou 2005). In 2012, an Organisation for Economic Co-operation and Development study concluded that actions to substantially reduce the amount of climate change are costly but well worth the investment (OECD 2012, 5–6).

The relationships between climate change and some major climate and weather events are better understood than they were a few years ago. The Fifth Assessment of global warming science from the Intergovernmental Panel on Climate Change found that in many parts of the world, "extreme precipitation events ... will *very likely* become more intense and more frequent" (emphasis in original). The panel said it was already likely that the frequency and duration of heat waves had increased and projected with greater than 90 percent confidence that heat waves will be more severe later in this century. While some research points to fewer hurricanes in a warmer climate, other studies find that the hurricanes that do form are likely to attain greater strength and have more associated rainfall (IPCC 2013, 5, 21; IPCC 2007, 751).

The ability of individual species to adapt to climate change varies widely. For example, a recent study compared bloom dates in 2010 and 2012 with data from more than 150 years ago for a group of flowers in Massachusetts and Wisconsin. Summarizing the study, one of the authors, Professor Charles Davis, reports, "We're seeing spring plants that are now flowering on average over three weeks earlier than when they were first observed—and some individual species that are flowering as much as six weeks earlier," adding that "these are far and away the earliest flowering times on record for the eastern United States." Professor Davis placed the findings of the study in the context of climate change. To illustrate, he said it is noticeably "warmer today than when Thoreau studied Concord. When we talk about climate change, it can be difficult to grasp what it means when we talk about future increases in temperature. Humans may weather these changes reasonably well in the short term, but many organisms in the tree of life will not likely fare nearly as well" (Reuell 2013).

A longer growing season does not always mean good news. According to federal data, "The warm spring resulted in an early start to the 2012 growing season in many places, which increased water demand on the soil earlier than what is typical. In combination with the lack of winter snow and lingering dryness from 2011, the record-warm spring laid the foundation for the great drought of 2012" (NOAA January 8, 2013).

Results from over 800 scientific papers on ecology offer a picture of the changes in species vitality and distribution: "People are finally starting to see the changes, spread across the world from the tropics to the Arctic and across every taxonomic group," notes ecologist Camille Parmesan of the University of Texas at Austin, who reviewed the behavior of 1,700 species of plants and animals (Hotz 2006, 1).

At one time, adapting to climate change—as opposed to acting now to decrease greenhouse gas emissions—was almost an ideological position. Refusal to discuss adapting to climate change was presented by some as denying scientific facts or avoiding problems because they were not convincingly presented or might cost too much and produce too little. In 2006, experts in the United Kingdom concluded: "An equitable international response to climate change must include action on both adaptation and mitigation. Adaptation and mitigation are not choices: substantial climate change is already inevitable over the next 30 years, so some adaptation is essential" (Stern 2006, 4). During international negotiations in Doha in 2012, the delegate from the Philippines broke down in tears as he urged the delegates to take action. In an interview with a reporter after his speech, the delegate explained: "We lose nearly 5 percent of our economy every year to storms. We have received no climate finance to adapt or to prepare ourselves for typhoons and other extreme weather we are now experiencing. ... So more and more people die every year ... it tears your heart out when you know your people are feeling the impact. We cannot go on like this. It cannot be a way of life that we end up running always from storms" (Vidal 2012).[1]

Experiences with Hurricane Sandy in 2012 and other recent extreme weather events suggest that governments should upgrade infrastructure now without waiting to see how other governments will respond.

As noted by Thompson and Bendik-Keymer, "problems like climate change involve global collective action and bring to the fore the way our individual lives are enmeshed in and constrained by institutions, including the ways we organize our lives and our politics" (2012, 11).

Through nongovernmental organizations, businesses, governments, and international communities, societies across the world are learning more and more about climate change. Rather than wait for greater certainty, many are moving forward at the local, state, and regional levels to reduce greenhouse gas emissions and strengthen resiliency to climate change, but more progress is needed. Overall, emissions from countries with Kyoto targets fell significantly, but global emissions have increased in the Kyoto target period. Writing in 2009, Ackerman notes, "If all of the US matched the performance of California and New York, the result would be a reduction of 40 percent of US emissions, or 8 percent of global emissions" (2009, 128–29).

In addition, opportunities to reduce greenhouse gas emissions on a voluntary basis are widely available. For example, there is a large market for voluntary greenhouse gas emission offsets. Tax credits and incentives encourage businesses and individuals to increase energy efficiency and use renewable energy.

New ideas are encouraged through competitions such as the Solar Decathlon, sponsored by the US Department of Energy. The MIT Center for Collective Intelligence has developed a project encouraging individuals to contribute ideas to reduce greenhouse gas emissions and increase resiliency to climate change. The project is called Climate CoLab.

In 2012–13, the Climate CoLab is dividing the overall problem of climate change into many different sub-problems (like how to reduce emissions from electric power generation in the largest emitting countries or how city governments can adapt to climate change). For each key sub-problem, experts in the area will advise community members as they develop proposals ... winning proposals will be presented to people and organizations who could ... implement them. (Climate CoLab 2013)

Although there is more to know about the effects of climate change and about the costs associated with various responses to it, this should not be a reason for delay: much is already known (Congressional Budget Office 2009).

The Economics of Climate Change and Risk Assessment

Although we need more information to fully understand the actual costs and benefits of meaningful responses to climate change, the benefits of reducing emissions without delay outweigh the costs. An expert in economic assessment of climate change finds that waiting fifty years to reduce carbon dioxide (CO_2) emissions would create a loss estimated to be $6.5 trillion (Nordhaus 2013, 300). Also, another expert economist argues for bold action at once: "If we spend twenty or thirty years talking about the need to get started and squabbling over shares of the costs, it will be all but impossible to avoid temperature increases that imply very dangerous climate risks" (Ackerman 2009, 127–128).

People who vary in their assessment of, tolerance for, and acceptance of risks treat the costs and risks associated with climate change differently. Numerous, noisy, and sometimes countervailing conclusions are reached by entities that influence public understanding. The media wax hot and cold about environmental issues, although the trend recently has been a strong set of communications that emphasize the hot. *Time, National Geographic, Rolling Stone,* the *New York Times, San Francisco Chronicle,* the *Los Angeles Times,* and the *New Yorker* have all published major articles on what they characterize as the serious nature of climate change. Nonetheless, the media in the United States seem structurally required to keep the issue as an ongoing debate. A few vocal and influential scientists remain contrarians and are given disproportionate attention in the media, in legislative committees, and in other policy forums. Business increasingly talks green, but consumers sometimes experience only a greenwash. Governments differ in how they state the problem, but few remain that refuse to acknowledge there is one.

There are different risks for different regions. The weakest and poorest communities will feel the harm most. But here there is another complicating factor: serious effects across regions—from migration, from movement of environmental health problems, and from the potential for political unrest that is linked to serious environmental degradation.

The meaning of climate change for us, our children, and our grandchildren will be influenced by whether we make investments today to reduce the amount of sea level rise, the frequency and severity of extreme weather events, the magnitude of environmental health challenges, and the strains on international relations for future generations. Some economic assessments of climate-change policies rely on cost-benefit analysis. In contrast, Ackerman (2009) recommends assessing the economics of climate-change policies as one would assess insurance against fire or other natural disasters.

The Function and Complexity of Cost-Benefit Analysis

Policy makers often use cost-benefit analysis to help select among competing options. For decades, economists have tried to assist policymakers in decisions by conducting cost-benefit analyses of proposed actions. The area of climate change is no exception. The Congressional Budget Office has reported,

Over the past 15 years, a large number of studies have analyzed the potential costs and benefits of averting climate change. Some researchers have attempted to incorporate the studies' results in global and regional models of economic growth and climate effects and have used models to conduct so-called integrated assessment of policy proposals related to climate change. They have also estimated the cost of emission control policies that would yield the greatest net benefits in terms of economic growth, reduced emissions, and the resulting climate effects. (2003, 27)

Rather than showing a range of possible conditions, some studies make a number of controversial assumptions (Congressional Budget Office): "technological change will probably lower the cost of controlling emissions," or "people are likely to be wealthier in the future," or "for carbon dioxide, emissions that occur sooner rather than later will have more time to be absorbed from the atmosphere by the oceans" (2003, 29). These are assertions worth debating and help create a context for making decisions on how to proceed.

To develop cost-benefit analyses for climate change, scientists and economists estimate socioeconomic costs of a range of environmental impacts predicted by climate models (computer simulations). Analysts also quantify costs of any proposed mitigation or adaptation measure, and then compare those costs for each option. This provides an estimate of whether the benefits of a proposed mitigation or adaptation measure exceed the economic costs for the conditions, scenarios, impacts, and discount rate assumed in the study. For example, a cost-benefit assessment was included in the California Air Resources Board's (California ARB) preparation of the cap-and-trade program for California's AB32 program and updated in 2010. The study found that such "policies can shift the driver of economic growth from polluting energy sources to clean energy and efficient technologies, with little or no economic penalty. These results are consistent with most other economic analyses of AB32 and of proposed federal climate-change legislation" (California ARB 2010). Views on the ARB study vary widely, as occurs with many cost-benefit analyses. Views differ on the ARB study's underlying assumptions, its alleged failure to incorporate certain variables, and its findings.

Another example of cost-benefit analysis is a study of the cost of damages from a severe hurricane compared to the cost of infrastructure upgrades and reinforcement designed to guard against such damage. For example, the cost of Hurricane Sandy could reach $50 billion, which may be five times greater than the annual costs of the most costly environmental regulations. Hurricane Katrina's costs were even higher: that storm resulted in an estimated $81 billion in damages (TodayNews-Gazette.com 2012), or approximately eight times more than the yearly cost of the costliest environmental regulations. These types of comparisons help to clarify the economic costs of failing to act.

Costs and benefits are also influenced by what economists call a *discount rate*. One dollar today is worth more than a dollar tomorrow; people place less value on the future than they do on the present. As the Congressional Budget Office points out, "at discount rates that approximate market rates, even very large long-term costs and benefits are dramatically devalued. ... The choice of discount rate therefore makes a huge difference in thinking about long-term problems such as climate change" (2003, 29).

For climate change, cost-benefit analysis is further complicated by scientific uncertainty. Although some climate change has already been set in motion, uncertainty remains regarding the amount and pace of temperature change and the magnitude of its effects. Moreover, the global, regional, and intergenerational nature of climate change renders the use of cost-benefit analysis even more difficult (Munasinghe and Swart 2005, 192). Finally, cost-benefit analyses are based on market values, which do not necessarily reflect social values or environmental justice values. Nor do these types of valuations give everyone an equal vote in determining the value of a specific action (Sandel 2012, 8, 31).

The EPA has concluded that acting to reduce greenhouse gas emissions results in significant economic benefits, ranging from reduced human health and welfare risks as a result of lower greenhouse gas emissions to lower levels of climate change and global warming. Whenever federal agencies issue regulations to implement statutes enacted by Congress, they must first justify any proposed regulations by conducting an analysis of the costs and benefits of the regulation to the economy and society. To analyze the economic benefits of EPA rules aimed at reducing CO_2 emissions, the EPA and other federal agencies have created social cost of carbon (SCC) estimates. SCC estimates help to calculate the economic damages associated with a specified increase in CO_2 emissions in any given year; this dollar figure also represents the value of damages avoided for a specific emission reduction. SCC estimates are used to represent the benefits used in cost-benefit analyses. When a proposed regulation would decrease emissions, the SCC represents the potential damages on a dollar-per-ton-of-CO_2-avoided basis. This calculation helps to determine the benefits associated with the proposed rule. However, current models cannot yet incorporate all of the important known or unknown physical, ecological, and economic impacts of increasing concentrations of CO_2 in the atmosphere due to a lack of precise data on the nature of the damages and because the scientific basis of the models does not integrate the latest findings in rapidly advancing climate change research. For these reasons, the benefit estimates are incomplete. Moreover, the SCC does not include the additional benefits that result from reductions in other pollutants due to regulations aimed at greenhouse gas emissions (EPA 2013a).

In an attempt to address these limitations, the US government plans to update the current estimates to reflect future improvements in climate science and economic understandings of climate change and its societal impacts, and has hosted a series of workshops on integrated assessment modeling, valuing climate change impacts, incorporating findings into policy analysis and estimating and valuing damages on a sectoral basis (EPA 2013b). In spite of its limitations, the SCC plays a key role in the assessment of CO_2-reduction benefits.

However, other assessments, dialogue, and political processes are used to inform decisions on who bears the costs and who reaps the benefits of public policies, including climate-change policies. "Policies that balance overall costs and benefits do not necessarily balance them for every person, and policies that maximize the net benefits to society do not necessarily provide benefits to each individual. A policy may yield positive net benefit by causing both very large aggregate losses and only slightly larger aggregate gains" (Congressional Budget Office 2003, 33). For climate change, the distribution of costs and benefits must be considered across economic groups, geographic regions, and generations. The international conferences of the parties (COPs) on climate change address this issue explicitly often and implicitly always. Debate continues over the roles that should be played by India, China, and other rapidly developing large countries in climate change. Although greenhouse gas emissions in these regions were relatively low in the past, they are now significant and are expected to continue to grow quickly in coming decades unless carbon-reducing strategies and technologies are deployed quickly.

Integrated assessments can also help, although they sometimes make controversial assumptions. A study by William Nordhaus is an example of how this technique can be used to analyze tradeoffs. The study finds that "the economic damages from climate change with no interventions will be on the order of 2.5 percent of world output per year by the end of the twenty-first century. The damages are likely to be most heavily concentrated in low-income and tropical regions such as tropical Africa and India" (2008, 6; see also 2013, 139).

Some of the economic questions on cost arise from speculations and best guesses about how governments and the private sector will implement programs to address climate change. Some programs have begun, and some have faltered. In March 2013, the greenhouse gas markets in the European Union and the eastern United States had prices that were very low; however, the California market was trading at higher prices. Although carbon markets provide funds for clean energy programs, the low carbon prices in some markets have prompted calls for a carbon tax to provide a better market signal for carbon reduction. At the same time, manufacturers in California are weighing how to respond to higher carbon prices while remaining competitive with businesses outside of the state.

Some costs are becoming more clear as we experience (rather than simply wonder about) climate change. We know where climate-change costs are likely to be felt most acutely, but predicted impacts at local or regional levels are by necessity very general. In agriculture, there will be increased costs (and, in places, benefits) associated with shifts in the type of crop cultivated and the water available for those crops. Industry, whether carbon dioxide is regulated directly or not, will need to pay for ground-level ozone noncompliance through stricter pollution controls or fines. Climate change is expected to increase ground-level ozone. A less healthy workforce (with increased asthma and other respiratory and heat-related illnesses) will take more sick days, decrease business productivity, and require higher health premiums. Higher temperatures will increase energy use, and peak demand will be higher in some regions unless measures are taken to further increase energy efficiency. Forest fires will be more frequent and more severe.

The estimates of the costs of reducing greenhouse gases vary with timescale and with understandings of the value we put on things that will happen to us or to our children and grandchildren in the future. The Congressional Budget Office summarized the situation blandly but accurately: "Climate policy thus involves balancing investments that may yield future climate-related benefits against other, non-climate related investments—such as education, the development of new technologies, and increases in the stock of physical capital—that are also beneficial. If climate change turned out to be relatively benign, a policy that restricted emissions at very high expense might divert funds from other investments that could have yielded higher returns. Conversely, if climate change proved to be a very serious problem, the same policy could yield a much higher return" (2003, 27).

The economics of climate change can work for society in more direct ways, including through many opportunities for the business sector. For example, the 2010 California Climate Action Team report to the state's governor described progress on a suite of strategies to increase investments in low-carbon technologies, including research, mitigation, adaptation, and joint action with regional and international efforts. The strategies include voluntary actions, incentives, and regulatory programs (California Environmental Protection Agency Climate Action Team 2010). The California Air Resources Board estimates the state's programs to reduce greenhouse gas emissions will create billions of dollars in savings through energy efficiency, attract billions of dollars of new investment, and bring hundreds of thousands of jobs to California (California ARB 2010). In 2013, Governor Jerry Brown took a large group of businesspeople in his delegation to China to promote California green technology as well as to encourage country-state cooperation on mitigation, demonstrating an example of opportunities to work across borders to address climate change.

Climate Change and Insurance Policies

Rather than focusing on the high-probability outcomes of climate change, Ackerman (2009) focuses on low-probability catastrophic events from climate change. Homeowners buy insurance against the risk of fire; although the risk of fire is low, the consequences would be unmanageable without insurance. Similarly, damages from low-probability, catastrophic risks of unmitigated climate change would be unmanageable. Ackerman argues that we should act without delay to reduce the chances of catastrophic climate change (Ackerman 2009, 30–42).

Ackerman points to the long-term investments taken by the Netherlands to guard against damages related to sea level rise as an example of "good costs" to prevent costly damages, loss of life, and loss of other priceless resources:

This may not be a perfect or permanent solution to the problems of rising seas and stormy weather, but it has successfully protected the country for many years. ... Would the Dutch have been better off building less extensive sea walls and barriers, taking a greater chance on damages from extreme storms in order to have even more money for private consumption? It is easier to believe the opposite: the long process of construction itself undoubtedly contributed to the economic growth of the Netherlands, providing employment and incomes for many workers for many years. (Ackerman 2009, 62)

Thinking about climate change in terms of insurance costs is starting to take root in the insurance industry. The insurance industry is beginning to consider the risks of climate change in its policies. Investment firms and funds can focus on companies that produce technologies and services that help mitigate or adapt to climate change. The California State Teachers Retirement System (CalSTRS), with more than $130 billion in assets, is working to improve corporations' climate-risk disclosure and response as part of its work on improving performance of its investment portfolio. CalSTRS was one of twenty large public and private US investors (collectively controlling more than $800 billion in assets) that pressured thirty large, publicly held insurance companies to disclose financial risks of climate change for life, health, and property insurance profitability. The investors also asked the insurance companies to identify steps they are taking to reduce exposure to these risks and explore new business opportunities in the changing economic environment expected from climate change. Insurance agencies that take climate into consideration also can develop new types of policies and cancel less profitable coverage. As of 2012, large insurance companies doing business in California, New York, and Washington states were required to report climate-related risks. Data for 2012 suggest that the insurance industry is not very well prepared for climate-related risks, although some are doing much better than others (CERES 2013).

In 2013, the Geneva Association, which conducts research on global insurance and risk management issues, published a report on the implications of warming oceans for the insurance and re-insurance industry. The report recommends the development of a new definition of "normal" weather conditions for use by insurers. It also recommends risk education to increase awareness of climate risks and support from the insurance industry for innovative, climate-resilient building stock (Geneva Association 2013; Cronin 2013).

The Issue in Context

What climate change means for you, your children, and your grandchildren depends on which generation you are considering, on where you live, and on how you value the risks and rewards of reducing greenhouse gas emissions. But as we have seen, it also depends on the actions that are taken by international organizations, nations, states and municipalities, regions and provinces, and businesses that directly influence climate policy. Citizens can also play a role in helping to mitigate and respond to climate change at home, in business, and through participation in government processes. Policymakers increasingly include climate change in their view of the world. Climate change is becoming part of the calculus on whether to switch to a low-carbon automobile fleet or personal car, to insure or refuse to insure, to invest in renewables or exploration of non-renewables, to minimize the environmental footprint of new homes or not (for those fortunate enough to have the choice): the list is almost as long as the sources of greenhouse gases.

A Final Word

Finally, we may address the almost statistically negligible possibility that we find climate change is either (a) not as harmful as we are virtually assured it will be, or (b) that its causes are largely unrelated to our actions related to carbon and other climate altering gases.

In such a setting, for the improvement of public health, we can better address the interactions of energy and climate policy with air quality (Bryner and Duffy 2012). And for the good of our grandchildren, for the equity associated with caring about the poor in Bangladesh, Tanzania, and other less-developed places, we might very well take the same path as one demanded by the consensual scientific conclusion: that our behaviors are changing the climate of a fragile planet (Matthew, this volume; Adger et al. 2006).

Whether one calls it *no regrets* or something else, for the good of those who are born in 2050 or 2070 or 2090, and those born today in low-lying and poor countries, we can act now to counter food shortages associated with drought, slow down the destruction of the great coral, maintain habitat of endangered species, find solutions to temperature increases in places already dreadfully hot, and work to protect those whose homes and livelihoods are close to the shorelines.

Note

1. As this book was going to press, Typhoon Haiyan, the strongest storm known to history, struck the Philippines. In addition to sustained winds of 195 mph and gusts of 235 mph, it produced an immense storm surge that many survivors thought was a tsunami. Several islands have been totally devastated. At this writing, the estimated death toll, which currently numbers in the thousands, and the economic impacts of the typhoon, are unknown.

References

Ackerman, Frank. 2009. *Can We Afford the Future? The Economics of a Warming World.* New York: Zed Books.

Adger, W. Neil, Jouni Paavola, Saleemul Huq, and M. J. Mace. 2006. *Fairness in Adaptation to Climate Change.* Cambridge, MA: MIT Press.

Baer, Paul, and Tom Athanasiou. 2005. Honesty about dangerous climate change. http://www.ecoequity.org/2005/11/honesty/#more-361, accessed July 7, 2013.

Bryner, Gary C., and Robert J. Duffy. 2012. *Integrating Climate, Energy, and Air Pollution Policies.* Cambridge, MA: MIT Press.

California Air Resources Board. 2010. *California's Climate Plan.* http://www.arb.ca.gov/cc/cleanenergy/clean_fs2.htm, accessed September 7, 2013.

California Environmental Protection Agency Climate Action Team. 2010. *Climate Action Team Report to Governor Schwarzenegger and the California Legislature.* Sacramento, CA: California Environmental Protection Agency.

CERES. 2013. *Is the U.S. Insurance Industry Prepared for Climate Change?* http://www.ceres.org/press/press-releases/is-the-u.s.-insurance-industry-prepared-for-climate-change, accessed September 7, 2013.

Climate CoLab. 2013. *About the Project.* http://climatecolab.org/about.

Congressional Budget Office. 2003. *The Economics of Climate Change: A Primer.* Washington, DC: Congressional Budget Office.

Congressional Budget Office. 2009. *Potential Impacts of Climate Change in the United States. A CBO Paper.* May. Washington, DC: Congressional Budget Office.

Cronin, John. 2013. Report: A "new normal" needed to protect insurance companies from climate change. Earthdesk Blog, July 5. Pace Academy for Applied Environmental Studies. Pace University. http://earthdesk.blogs.pace.edu/2013/07/05/a-new-normal-to-protect-the-insurance-industry-from-climate-change/, accessed July 5, 2013.

Environmental Protection Agency. 2013a. The Social Cost of Carbon. http://www.epa.gov/climatechange/EPAactivities/economics/scc.html, accessed September 7, 2013.

Environmental Protection Agency. 2013b. Fact Sheet: Social Cost of Carbon. July 2013. http://www.epa.gov/climatechange/Downloads/EPAactivities/scc-fact-sheet.pdf, accessed September 7, 2013.

Geneva Association. 2013. *Warming of the Oceans and Implications for the (Re)insurance Industry.*

Hotz, Robert Lee. 2006. Tapping into a changing climate. *Los Angeles Times,* 32, A1.

Hulme, Mike. 2010. The year climate science was redefined. http://www.theguardian.com/environment/2010/nov/15/year-climate-science-was-redefined, accessed September 7, 2013.

Intergovernmental Panel on Climate Change. 2007. *Climate Change 2007: The Physical Science Basis.* Contribution of Working Group I to the Fourth Assessment Report of the Intergovernmental Panel on Climate Change. Edited by S. Solomon, D. Qin, M. Manning, Z. Chen, M. Marquis, K. B. Averyt, M. Tignor, and H. L. Miller. Cambridge, UK and New York: Cambridge University Press.

Intergovernmental Panel on Climate Change. 2013. Summary for Policymakers. In Climate Change 2013: The Physical Science Basis. Contribution of Working Group I to the Fifth Assessment Report of the Intergovernmental Panel on Climate Change. Edited by T. F. Stocker, D. Qin, G.-K. Plattner, M. Tignor, S. K. Allen, J. Boschung, A. Nauels, Y. Xia, V. Bex, and P. M. Midgley. Cambridge, UK and New York: Cambridge University Press.

Munasinghe, Mohan, and Rob Swart 2005. *Primer on Climate Change and Sustainable Development: Facts, Policy Analysis and Applications.* Cambridge, UK: Cambridge University Press.

Nordhaus, William D. 2008. *A Question of Balance. Weighing the Options on Global Warming Policies.* New Haven, CT: Yale University Press.

Nordhaus, William D. 2013. *The Climate Casino: Risk, Uncertainty, and Economics for a Warming World.* New Haven, CT: Yale University Press.

OECD. 2012. *OECD Environmental Outlook to 2050: The Consequences of Inaction.* Highlights. http://www.oecd.org/env/indicators-modelling-outlooks/49846090.pdf.

Reuell, Peter. 2013. An early sign of spring, earlier than ever. http://news.harvard.edu/gazette/story/2013/01/an-early-sign-of-spring-earlier-than-ever/.

Sandel, Michael J. 2012. *What Money Can't Buy: The Moral Limits of Markets.* New York: Farrar, Straus and Giroux.

Stern, Sir Nicholas. 2006. What is the economics of climate change? *World Economy* 7 (2):1–10.

Thompson, Allen, and Jeremy Bendik-Keymer. 2012. Introduction: Adapting humanity. In *Ethical Adaptation to Climate Change: Human Virtues of the Future*, edited by Allen Thompson and Jeremy Bendik-Keymer, 1–23. Cambridge, MA: MIT Press.

TodayNewsGazette.com. 2012. Hurricane Katrina $81.2 Billion in Damage. Published October 30 by staff. http://todaynewsgazette .com/hurricane-katrina-81-2-billion-in-damage/, accessed September 7, 2013.

Vidal, John. 2012. Will Philippines negotiator's tears change our course on climate change? *The Guardian*. December 6.

Glossary

abrupt climate change   A large, rapid, unexpected change in average weather conditions that affects local, regional, and global patterns.

acid rain or precipitation   The removal of acidic gases and *aerosols* from the atmosphere to the earth's surface by fog, dew, rain, and snow. The main source of acid-forming gases is the release of sulfur dioxide and nitrogen from the burning of fossil fuels.

acidification   Reduction of the pH of seawater caused by increased absorption of CO_2 from the atmosphere.

adaptive capacity   In *Climate Change 2007: Impacts, Adaptation and Vulnerability. Contribution of Working Group II to the Fourth Assessment Report of the Intergovernmental Panel on Climate Change*, ed. M. L. Parry et al. (Cambridge, UK: Cambridge University Press, 2007), the IPCC defines adaptive capacity as "the ability or potential of a system to respond successfully to climate variability and change, and includes adjustments in both behaviour and in resources and technologies" (727).

aerosols   Suspended solid and liquid particles that range in size from tiny molecular clusters to particles visible to the human eye. Some are man-made, spewed by smokestacks, unfiltered tailpipes, and volcanoes. They can have both cooling and warming influences on the *atmosphere*.

albedo   The reflectivity of a material. According to the *Intergovernmental Panel on Climate Change*, albedo is the fraction of solar radiation reflected by the surface or object, often described as a percentage.

anthropogenic   Human-caused or -related.

atmosphere   The envelope of gases that surrounds the earth or other planets.

atmospheric circulation   Atmospheric motion that is a response to the unequal heating of the earth's surface; redistribution that acts to create energy balance.

biochar   A solid material created by heating biomass in an environment low in oxygen. Commonly used as a soil amendment, biochar sequesters, or removes, carbon from the atmosphere.

biomass   Material and waste products from living or recently living plants, animals, or other organisms.

boreal forest   An evergreen coniferous forest of the subarctic and regions with cold continental climate dominated by firs and spruces.

biosphere   The part of the earth system that includes all *ecosystems* and living things on land, in the *atmosphere,* and in the oceans.

cap and trade   A program in which an entity, such as a business, has a limit on a substance, such as *greenhouse gases,* that it can emit but is allowed to buy credits from other entities if it exceeds its limits or sell its credits if it emits less than its cap.

carbon   An abundant chemical element essential for life. In the form of *carbon dioxide (CO$_2$)* in the atmosphere, it acts as a *greenhouse gas,* raising the temperature of the earth. See also *carbon dioxide.*

carbon cycle   The exchange of *carbon* among the atmosphere, oceans, and land, including the *biosphere* and *lithosphere.*

carbon dioxide (CO$_2$)   A heavy, colorless, nonflammable gas that makes up about 0.04 percent of the earth's *atmosphere.* The amount of *carbon dioxide* oscillates with the seasons as deciduous trees and plants increase the amount they draw out of the air during their growing season, using it to build biomass. The amount of *carbon dioxide* in the atmosphere has increased in recent decades beyond the largest amount known in the past hundreds of thousands of years.

carbon dioxide equivalence   A measure that compares the radiative *forcing* of a given *greenhouse gas* to that of *carbon dioxide. Forcing* refers to the difference of radiant energy received by the earth and that radiated back. See also *forcing* below.

carbon sequestration   The long-term storage, through natural processes or technology, of *carbon* in the terrestrial *biosphere,* underground, or in the oceans. The aim is to reduce or slow the buildup of *carbon dioxide* in the atmosphere.

carbon trading   An exchange mechanism for trading *carbon* emission rights as a means of meeting emission-reduction targets. Those who trade in carbon buy and sell a type of contract that allows them to emit, reduce emissions, or offset against emissions.

chaos theory   The theory that minute inaccuracies in initial observations (such as temperatures or winds, in the climate-change case), when placed into a forecast model, will grow exponentially in time. It is also referred to as the *butterfly effect,* after the idea that the force of a butterfly flapping its wings in Brazil could cascade and set off a tornado in Texas.

chlorofluorocarbons (CFCs) and related chemicals   Synthetic chemicals that are odorless, nontoxic, nonflammable, and chemically inert; generally used for refrigerants, fire retardants, and aerosol sprays, they cause a seasonal decrease in the amount of protective ozone in the *stratosphere.*

Clean Development Mechanism   A device provided by the *Kyoto Protocol* to allow developed countries to earn emission-reduction credits for investing in eligible projects located in developing countries. To be eligible, projects must make greenhouse gas emission reductions that are "real, measurable, verifiable and additional to what would have occurred without the project" and meet other requirements. For further information, see the United Nations Framework Convention on Climate Change, 2007, *Kyoto Protocol Mechanisms*, available online at http://unfccc.int/resource/docs/publications/mechanisms.pdf.

climactic niche   Area providing acceptable conditions for a species to persist and reproduce.

climate   An average of *weather* conditions over an extended period of time (months, years, or centuries). The *World Meteorological Organization* uses a thirty-year period. Climate is characterized by temperature, precipitation, cloud cover, humidity, and wind patterns.

climate change   A significant change from one climatic condition to another. The term often refers to the buildup of man-made gases in the *atmosphere* that trap the sun's heat, causing changes in global *weather* patterns. Some use the term interchangeably with *global warming;* many scientists use the term to include natural changes in climate.

climate forcing   See *forcing*.

computer modeling and climate models   The major control system for studying the complexity of the earth system. Because scientists cannot conduct classical experiments to study climate change, they use models of varying and increasing complexity. The *Intergovernmental Panel on Climate Change* defines a *climate model* as a numerical representation of the *climate* system that is based on the physical, chemical, and biological properties of its components, of their interactions, and of their feedback processes and that accounts for all or some of their known properties.

conference of the parties (COP)   The supreme decision-making body of the United Nations *Framework Convention on Climate Change (FCCC)*. It is composed of all of the nations that have ratified the agreement. Its first session was held in Berlin in 1995, and it has met nineteen times thereafter in The Hague, Milan, Montreal, and other cities. The nineteenth conference is scheduled for November of 2013.

consilience   The coming together of inductions. It occurs when one class of facts coincides with an induction from a different class.

cryosphere   The frozen water—snow, ice, and permafrost—on and beneath the surface of the earth. It includes the Greenland and Antarctic ice sheets and sea ice in the Arctic and Southern Oceans. It represents about 1 percent of the water on the planet.

deduction   The process of drawing logical inferences from a set of premises.

ecosystem   An interactive community of living organisms and their physical environment.

El Niño Southern Oscillation   An *atmosphere* and ocean phenomenon that channels year-to-year fluctuations in ocean temperatures over the tropical east Pacific into global fluctuations in *climate*.

emissions trading   A market-based system allowing those who can reduce emissions at a lower cost to sell their spare emission units (allowances) and emission-reduction credits (offsets) at a profit to others. Buyers use the allowances and credits to offset their own emissions (such as using fossil-fuel powered electricity generation) or emissions of the goods or services they consume (such as running a carbon-neutral business).

ensemble   A large collection of models used to explore diverse possible outcomes.

equilibrium A stable state in a dynamic system in which changes are balanced by other changes.

error bars Used in statistics, these are the range of values estimated to contain the actual value area within a specified level of confidence uncertainty, usually 95 percent.

falsification The position in the study of scientific thinking that holds a theory can never be proven true but it can be proven false.

flexibility mechanisms Devices for fostering implementation of the *Kyoto Protocol*'s target for reducing *greenhouse gas* emissions while limiting costs. These include *joint implementation,* the *clean development mechanism,* and *emissions trading.*

forcing Any imposed mechanism that forces *climate* to change. Natural forcing results from volcanic eruptions and solar variability; man-made or *anthropogenic* forcing comes from the use of *greenhouse gases* and *aerosols.* In *Climate Change 2001: The Scientific Basis,* ed. J. T. Houghton et al. (Cambridge UK: Cambridge University Press, 2001), the IPCC used the term "*radiative forcing*" to describe an "externally imposed perturbation in the radiative energy budget of the Earth's climate system" (353). Also, see *radiative forcing* below.

Framework Convention on Climate Change (FCCC) An agreement signed in 1992 at the United Nations Conference on Environment and Development that set out general parameters and principles for international efforts to address *climate change,* with the aim of stabilizing *greenhouse gas* concentrations in the *atmosphere* at a level that would prevent dangerous *anthropogenic* interference with the climate system. The agreement went into effect in 1994.

geoengineering Large-scale manipulation of an environmental process of the earth designed to achieve a preferred result. Some *geoengineering* ideas have been proposed that could, hypoethically, modify or lessen the effects of climate change. However, there are many uncertainties, risks, and concerns related to such proposals.

Global Climate Change Initiative The US program, started in the George W. Bush administration, that aims to reduce the ratio of *greenhouse gas* emissions to economic output through domestic voluntary actions and continued research on *climate change.*

Global Climate Coalition A group of major national environmental organizations whose goal is to influence the official US response to *climate change.*

global warming   See *climate change.*

global warming potential (GWP)   An index that estimates the heat-trapping efficiency of *greenhouse gases.* It represents the ratio of energy reemitted to the earth's surface during a year for a given gas compared to that of the same mass of carbon dioxide.

greenhouse effect   The warming of the planet because of the existence of the earth's *atmosphere.* The effect is similar to how a plant warms when it is encased in a house of glass or how a blanket traps body heat. It provides that the average surface temperature of the earth warms to 15°C (59°F). Greenhouse gases absorb thermal radiation emitted from the earth's surface and then reradiate this energy back to the surface of the earth—allowing temperatures to be significantly warmer than they would be in the absence of an atmosphere.

greenhouse gas   Certain trace gases in the earth's *atmosphere* that selectively absorb and trap longer wavelengths (infrared radiation) of energy emitted by the earth and reemit them back to the earth's surface, allowing for a significant warming of the earth's surface and of the lower atmosphere. They account for less than 3 percent of the atmosphere and include *carbon dioxide (CO_2), methane (CH_4), nitrous oxide (N_2O), halocarbons, ozone (O_3), water vapor (H_2O), perfluorinated carbons (PFCs),* and *hydrofluorocarbons (HFCs).*

greenhouse gas efficiency   A measure of the ability of gases to absorb thermal radiation emitted by the earth; it is a function of the molecular structure of the gas.

greenhouse gas performance standard   A maximum allowable rate of *greenhouse gas* emissions from electricity generation. In California, the standard will be no lower than levels achieved by a new combined-cycle natural gas turbine. As of mid-2007, electric utilities in California are prohibited from entering into a long-term financial commitment (ownership or a contract of five years or more) for often-used power plants that exceed the greenhouse gas–emission performance standard.

gross domestic product   A measure of the economic activity that takes place within a country's borders during a period of time, calculated as consumption + investment + government spending + exports – imports.

Hadley circulation   An atmospheric circulation caused by warm air rising from the tropics to the upper troposphere and moving poleward while cold dense air near the poles sinks to the surface and moves toward the equator.

halocarbons Potent *greenhouse gases* that do not exist in nature. They include *hydrofluorocarbons* and *perfluorinated carbons.*

hydrofluorocarbons (HFCs) Alternatives to *chlorofluorocarbons* that were introduced in response to the *Montreal Protocol.* They do not destroy ozone, but they are a potent greenhouse gas, although with less *global warming potential* than CFCs.

hydrologic cycle The exchange of water among the atmosphere, land, and oceans through processes including precipitation and evaporation.

hypothetico-deductive model An approach often associated with the scientific method wherein scientists develop hypotheses and then test them.

induction The process of generalizing from specific examples.

integrated assessment An analysis of the advantages and disadvantages of adopting a particular action. Assessments involve literature reviews, quantification of losses associated with an action or failure to act, and cost-benefit analyses of effects on current generations against those on future generations.

Intergovernmental Panel on Climate Change (IPCC) The main international body established in 1988 by the *World Meteorological Organization* and the United Nations Environmental Programme to assess *climate-change* science and provide advice to the international community. The IPCC is an international group of scientists who summarize the current understanding of *climate change* and predict how *climate* may evolve.

joint development Refers to one of the general concepts underlying "flexibility" mechanisms for promoting implementation of the Kyoto regime with an aim of improving compliance. Flexibility mechanisms include *Joint Implementation,* the *Clean Development Mechansim,* and *emissions trading* systems. *Joint Implementation* enables industrialized countries to carry out projects jointly with other developed countries. The *Clean Development Mechanism* involves investment in sustainable development projects that reduce emissions in developing countries. *Emissions trading* systems allow countries that have spare emission units (allowances) and emission-reduction credits (offsets) to sell excess allowances or offsets to countries that are not meeting their targets.

Joint Implementation   One of the *flexibility mechanisms* defined in article 6 of the *Kyoto Protocol*. To meet their commitments, developed countries may exchange emission-reduction credits based on projects aimed at reducing emissions at sources or increasing removals by sinks of *greenhouse gases*.

Kyoto Protocol   The 1997 agreement that strengthened the United Nations *Framework Convention on Climate Change*. It committed developed countries to individual, legally binding targets to limit or reduce their *greenhouse gas* emissions. Individual targets for reducing *greenhouse gas* emissions for developed countries are listed in an annex to the Kyoto Protocol. These add up to a total cut in greenhouse gas emissions of at least 5 percent from 1990 levels in the commitment period 2008–2012. The Kyoto Protocol was revised in 2012 to add a second commitment period of 2013–2020. As of September 2013, the second commitment period is not yet in force. The Kyoto Protocol entered into force in February 2005, and as of late 2006, 166 states and regional organizations had ratified it.

lithosphere   The earth's crust and solid portion of the mantle.

Little Ice Age   A relatively minor *climate* variation in the fifteenth to the nineteenth centuries that was characterized by temperatures that were 2°C (3.6°F) cooler than today in much of Europe.

longwave or thermal radiation   Energy emitted at longer *wave lengths*, such as from the earth.

Medieval Warm Period   A relatively minor *climate* variation in the tenth to fourteenth centuries that was characterized by unusually warm weather in Europe.

meeting of the parties (MOP)   A meeting of the parties to the United Nations *Framework Convention on Climate Change* that occurred after the entry into force of the *Kyoto Protocol*. At Montreal in 2005, the first meeting of the parties ran parallel to the *conference of the parties* to the Convention.

methane (CH4)   A *greenhouse gas* that is odorless, colorless, and flammable. Among its sources is animal waste.

model calibration   Adjustments in the parameters of a model in attempts to better reproduce past data.

Montreal Protocol on Substances That Deplete the Ozone Layer An international agreement designed to protect the *ozone* layer by phasing out the production of several substances believed to be responsible for *ozone depletion*. Known as ozone-depleting substances, they are linked to what is commonly known as the *ozone hole*.

National Oceanic and Atmospheric Administration (NOAA) A federal agency within the US Department of Commerce that focuses on the condition of the oceans and the *atmosphere*.

network Interconnected people or things that may be used to communicate information, allocate access to farming or fishing resources, or transport people, goods, or services.

Nitrous oxide (N_2O) A *greenhouse gas* with a *global warming potential* of about 310. Sources include the burning of biomass, soil-cultivation practices, fertilizer use, fossil-fuel combustion, and nitric acid production.

no-regrets policy A public policy that favors options that have negative net costs and that generate both direct and indirect benefits that are large enough to offset the costs of implementation.

oceanic conveyor belt Overall ocean circulation. The three-dimensional ocean movement that encompasses the planet is a theoretical route driven by wind and by the *thermohaline* circulation.

Organization of Economic Cooperation and Development (OECD) An international body composed of thirty countries that addresses social issues and provides statistics. It facilitates multilateral agreements and aid to promote good governance and further economic growth.

ozone Three oxygen atoms combined into a molecule, which forms a protective layer in the *stratosphere* that limits the amount of *ultraviolet light* reaching the surface of the earth to generally healthful levels.

ozone depletion The reduction of protective *ozone* in the earth's *stratosphere*. Ozone depletion allows increased levels of *ultraviolet light* to reach the Earth's surface, increasing the risk of skin cancer. Ozone has been reduced through chemical reactions flowing from the release of *chlorofluorocarbons* and other ozone-depleting substances produced by humans.

paleoclimatology The study of past or ancient climates.

paleorecords   In the context of earth science, any record of the earth obtained from centuries past. These include records available from tree rings, ice cores, fossils, and sediment cores, among other sources.

perfluorinated carbons (PFCs)   Chemical compounds used in aluminum manufacturing, semiconductor manufacturing, and refrigeration. They do not harm stratospheric *ozone,* but they are potent *greenhouse gases.* PFCs are regulated by the *Kyoto Protocol.*

photosynthesis   The process plants (and some bacteria) use to make carbohydrates from CO_2 (if terrestrial) or bicarbonate, HCO_3, (if aquatic) in the presence of light and water. O_2 is released as a part of this process.

polar amplification   The tendency of large temperature changes to be much greater at the North or South Pole than the global average. This is thought to be caused by changes in the area covered by ice and snow, which absorb less energy from the sun than other surfaces.

policy entrepreneur   Individuals with deep knowledge on a topic who can influence policy outcomes because their expertise is not seen to be politically motivated.

positive feedback   A change in one element or variable that provides a change in a system.

proxy indicator   Substitutes for what is observable in nature. For climate changes, proxy data are natural recording systems of past climate found in sediments, ice cores, tree rings, and corals.

radiative forcing   An externally imposed change of the electromagnetic energy in the Earth's climate system. Drawing on work by V. Ramaswamy and colleagues, *Climate Change 2007: The Physical Science Basis. Contribution of Working Group I to the Fourth Assessment Report of the Intergovernmental Panel on Climate Change,* ed. S. Solomon et al. (New York: Cambridge University Press, 2007), uses the following definition: "the change in net (down minus up) irradiance … at the tropopause after allowing for stratospheric temperatures to readjust to radiative equilibrium, but with surface and tropospheric temperatures and state held fixed at the unperturbed values" (133).

salinity   The salt content of a substance such as water or soil measured as the percentage of dissolved salts by weight. It is also expressed in parts per million (ppm). Ocean salinity is also measured in practical salinity units (PSU). The salinity of the ocean varies by depth and location but is generally about 3.5 percent salt (35,000 ppm or 35 PSU).

scientific method   The process of conducting scientific inquiry. There is no universally accepted scientific method, but there is some consensus about the standards and criteria for judging what makes scientific information a reliable basis for action.

shortwave radiation   Energy emitted at short *wavelengths*, such as from the sun.

sink   Any process that removes *greenhouse gases* or their precursors and *aerosols* from the *atmosphere*.

smog   Visibly polluted air, historically caused by sulfur dioxide from burning coal mixed with fog. Since the widespread use of automobiles, *smog* also means photochemical smog, created by chemical reactions of sunlight, emissions from mobile and stationary fossil-fuel combustion, and vapors from widely used chemicals such as gasoline, industrial solvents, and pesticides.

stratosphere   The region of the *atmosphere* above the *troposphere* extending roughly from about 10 kilometers (or roughly 6 miles) to about 50 kilometers.

thermohaline circulation   Density-driven circulation in the ocean on a large scale. It is caused by differences in *salinity* and temperature.

toxic waste   Discarded gaseous, liquid, or solid substances that can cause death or injury to humans or the environment. In the United States, acute toxicity is measured by the amount of the substance that is lethal to 50 percent of test rodents after fourteen days.

troposphere   The part of the earth's *atmosphere* from the ground level up to at least 5 miles above the earth's surface. In some places, the troposphere extends to 9 miles above the earth's surface.

ultraviolet light   Energy with wavelengths slightly shorter than visible light.

water density   Weight per volume of water, which increases with salt, cold, and pressure. If water becomes salty, it becomes more dense and sinks below fresher water. Likewise, cold water is denser than warm water and sinks below warmer water.

water vapor   Water in a gaseous state.

wavelength   The distance between adjacent crests in the wave.

weather   The atmospheric conditions for an individual event or time. The current limit of the forecast of weather extends to about ten days.

World Meteorological Organization (WMO)   A United Nations agency concerned with the international collection of meteorological data.

Younger Dryas   A period nearly thirteen thousand years ago when glacial conditions rapidly returned to the high latitudes of North America and temperatures in northern Europe dropped nearly 5°C (9°F).

Contributors

John T. Abatzoglou, Ph.D., is an Assistant Professor in the Department of Geography at the University of Idaho. His research focuses on regional climate dynamics and climate impacts specific to western North America.

Joseph F. C. DiMento, Ph.D., J.D., is professor of Law and of Planning, Policy, and Design at the University of California, Irvine, and a board member of the Newkirk Center for Science and Society. He is the author or editor of a dozen books, including *Environmental Governance of the Great Seas: Law and Effect* with Alexis Hickman (2012); *Changing Lanes: Visions and Histories of Urban Freeways* with Cliff Ellis (2013); *The Global Environment and International Law* (2003), and *Environmental Law and American Business: Dilemmas of Compliance* (1986). He has contributed to these topics in the legal, social science, and policy literature and in the popular press, and he is a regular observer of climate-change negotiations.

Pamela Doughman, Ph.D., is an energy specialist at the California Energy Commission. Publications include "Water Cooperation in the U.S.-Mexico Border Region," in Ken Conca and Geoffrey D. Dabelko, eds., *Environmental Peacemaking* (2002); and "Discourse and Water in the U.S.-Mexico Border

Region," in Joachim Blatter and Helen Ingram, eds., *Reflections on Water: New Approaches to Transboundary Conflicts and Cooperation* (2001). She has contributed to a number of publications of the California Energy Commission.

Crystal A. Kolden, Ph.D., is an assistant professor in the department of geography at the University of Idaho. Her research interests include wildfire management, ecology, climate impacts, geographic information systems, remote sensing, and invasive species.

Suzanne M. Levesque, Ph.D., received her Ph.D. from the University of California, Irvine. Her publications include "The Yellowstone to Yukon Conservation Initiative: Reconstructing Boundaries, Biodiversity, and Beliefs" in Joachim Blatter and Helen Ingram, eds., *Reflections on Water: New Approaches to Transboundary Conflicts and Cooperation* (2001), and "Lessons in Transboundary Resource Management from Ambos Nogales" (with Helen Ingram) in Lisa Fernandez and Richard T. Carson, eds., *Both Sides of the Border: Transboundary Environmental Management Issues Facing Mexico and the United States* (2002). She currently works as a manuscript proofreader and editor. *Changing Lanes: Visions and Histories of Urban Freeways* (2012), authored by Joseph F. C. DiMento and Cliff Ellis, is among her most recent endeavors.

Richard A. Matthew, Ph.D., is professor of international and environmental politics in the Schools of Social Ecology and Social Science at the University of California, Irvine, and director of the Center for Unconventional Security Affairs (www.cusa.uci.edu) at UC Irvine. He is also a Senior Fellow at the International Institute for Sustainable Development in Geneva; a Senior Fellow at the Munk School at the University of Toronto; a senior member of the United Nations Expert Advisory Group on Environment, Conflict and Peacebuilding; and a member of the World Conservation Union's Commission on Environmental, Economic and Social Policy. Dr. Matthew has received Certificates of Recognition for his research and service activities from the US Congress, the California State Legislature, and the City of Los Angeles. He has over 150 publications, including seven books and coedited volumes.

Stefano Nespor, lecturer at the Polytechnic University and an attorney in Milan, is a leading environmental, labor, and administrative lawyer in Europe and editor of *Rivista Giuridica dell'Ambiente.* He is the author of numerous books on the environment and law, including a recently published book with Alda L. De Cesaris *(Le Lunghe estati calde. Il cambiamento climatico e il protocollo di Kyoto,* 2004) in Italian for the nonscientist on climate change.

Naomi Oreskes, Ph.D., is a professor of the history of science at Harvard University. Her research focuses on the historical development of scientific knowledge, methods, and practices in the earth and environmental sciences. One of her books, *Plate Tectonics: An Insider's History of the Modern Theory of the Earth* (2001), was cited by *Library Journal* as one of the best science and technology books of 2002. Her book, *Merchants of Doubt: How a Handful of Scientists Obscured the Truth on Issues from Tobacco to Global Warming* (2010), coauthored with Erik M. Conway, received the 2011 Watson Davis and Helen Miles Davis Prize for best book for a general audience.

Andrew C. Revkin, who has spent three decades covering science and the environment, is the Senior Fellow for Environmental Understanding at Pace University's Pace Academy for Applied Environmental Studies and writes the award-winning Dot Earth blog for the opinion section of the *New York Times,* where he was a reporter from 1995 through 2009. At Pace, he teaches courses on blogging, environmental science communication, and documentary video, with a focus on sustainable development. He has written acclaimed books on global warming, the changing Arctic, and the fight to save the Amazon rain forest and three book chapters on communication and the environment. His quarter-century of coverage of global warming has earned most of the major awards for science journalism, including an unprecedented pair of communication awards from the National Academy of Sciences.

Index

Note: italicized numbers refer to figures, tables, and boxes.

American and Comparative Environmental Policy

Sheldon Kamieniecki and Michael E. Kraft, series editors

William Ascher, Toddi Steelman, and Robert Healy, *Knowledge in the Environmental Policy Process: Re-Imagining the Boundaries of Science and Politics*

Michael E. Kraft, Mark Stephan, and Troy D. Abel, *Coming Clean: Information Disclosure and Environmental Performance*

Paul F. Steinberg and Stacy D. VanDeveer, editor, *Comparative Environmental Politics: Theory, Practice, and Prospects*

Judith A. Layzer, *Open for Business: Conservatives' Opposition to Environmental Regulation*

Kent Portney, *Taking Sustainable Cities Seriously: Economic Development, the Environment, and Quality of Life in American Cities*, second edition

Raul Lejano, Mrill Ingram, and Helen Ingram, *The Power of Narrative in Environmental Networks*

Christopher McGrory Klyza and David J. Sousa, *American Environmental Policy: Beyond Gridlock*, updated and expanded edition

Andreas Duit, editor, *State and Environment: The Comparative Study of Environmental Governance*

Joseph F. C. DiMento and Pamela Doughman, editors, *Climate Change: What It Means for Us, Our Children, and Our Grandchildren*, second edition